ASAP
Chemistry

By the Staff of The Princeton Review

princetonreview.com

Penguin
Random
House

The Princeton Review
110 East 42nd Street, 7th Floor
New York, NY 10017
Email: editorialsupport@review.com

Copyright © 2018 by TPR Education IP Holdings, LLC. All rights reserved.

Published in the United States by Penguin Random House LLC, New York, and in Canada by Random House of Canada, a division of Penguin Random House Ltd., Toronto.

Terms of Service: The Princeton Review Online Companion Tools ("Student Tools") for retail books are available for only the two most recent editions of that book. Student Tools may be activated only twice per eligible book purchased for two consecutive 12-month periods, for a total of 24 months of access. Activation of Student Tools more than twice per book is in direct violation of these Terms of Service and may result in discontinuation of access to Student Tools Services.

ISBN: 978-0-525-56767-7
eBook ISBN: 978-0-525-56778-3
ISSN: 2576-5809

AP is a trademark registered and owned by the College Board, which is not affiliated with, and does not endorse, this product.

The Princeton Review is not affiliated with Princeton University.

Editor: Selena Coppock
Production Editors: Harmony Quiroz and Liz Dacey
Production Artist: Deborah A. Weber

Printed in the United States of America.

10 9 8 7 6 5 4 3 2 1

Editorial
Rob Franek, Editor-in-Chief
Mary Beth Garrick, Executive Director of Production
Craig Patches, Production Design Manager
Selena Coppock, Managing Editor
Meave Shelton, Senior Editor
Colleen Day, Editor
Sarah Litt, Editor
Aaron Riccio, Editor
Orion McBean, Associate Editor

Penguin Random House Publishing Team
Tom Russell, VP, Publisher
Alison Stoltzfus, Publishing Director
Amanda Yee, Associate Managing Editor
Ellen Reed, Production Manager
Suzanne Lee, Designer

Acknowledgments

The editor of this book would like to thank the tremendous content development team: Nick Leonardi, Eliz Markowitz, Merissa Remus, Catherine Chow, and Suzy Drurey. Thank you so much for your hard work, imagination, enthusiasm, and expertise.

The reason that this series looks so gorgeous is because Debbie Weber and Craig Patches are on the case. Their unending dedication, hard work, and imagination have made working on this series an absolute pleasure. Is there anything cooler than asking for a smiley face icon and getting one who is wearing sunglasses?

Contents

Get More (Free) Content vi

Introduction ix

Chapter 1: Atomic Structure 1

Chapter 2: Covalent Bonding and Intermolecular Forces 25

Chapter 3: Stoichiometry, Precipitation Reactions, and Gas Laws 65

Chapter 4: Thermochemistry 105

Chapter 5: Equilibrium and the Solubility Product Constant 127

Chapter 6: Redox Reactions and Electrochemistry 151

Chapter 7: Acids and Bases 175

Chapter 8: Kinetics 199

Get More (Free) Content

1 Go to **PrincetonReview.com/cracking**

2 Enter the following ISBN for your book: 9780525567677

3 Answer a few simple questions to set up an exclusive Princeton Review account. (If you already have one, you can just log in.)

4 Click the "Student Tools" button, also found under "My Account" from the top toolbar. You're all set to access your bonus content!

Need to report a potential **content** issue?

Contact **EditorialSupport@review.com**.
Include:
- full title of the book
- ISBN
- page number

Need to report a **technical** issue?

Contact **TPRStudentTech@review.com**
and provide:
- your full name
- email address used to register the book
- full book title and ISBN
- computer OS (Mac/PC) and browser (Firefox, Safari, etc.)

Once you've registered, you can...

- Get valuable advice about the college application process, including tips for writing a great essay and where to apply for financial aid

- Use our searchable rankings of *The Best 384 Colleges,* if you are still choosing between colleges, in order to find out more information about your dream school

- Check to see if there have been any corrections or updates to this edition

- Get our take on any recent or pending updates to the AP Chemistry Exam

Introduction

What is This Book and When Should I Use It?

Welcome to *ASAP Chemistry*, your quick-review study guide for the AP Exam written by the Staff of The Princeton Review. This is a brand-new series custom built for crammers, visual learners, and any student doing high-level AP concept review. As you read through this book, you will notice that there aren't any practice tests, end-of-chapter drills, and there are only a handful of practice questions. There's also very little test-taking strategy presented in here. Both of those things (practice and strategy) can be found in The Princeton Review's other top-notch AP series—*Cracking*. So if you need a deep dive into AP Chemistry, check out *Cracking the AP Chemistry Exam* at your local bookstore.

ASAP Chemistry is our fast track to understanding the material—like a fantastic set of class notes. We present the most important information that you MUST know (or should know or don't need to know—more on that later) in visually friendly formats such as charts, graphs, and maps, and we even threw a few jokes in there to keep things interesting.

Use this book any time you want—it's never too late to do some studying (nor is it ever too early). It's small, so you can take it with you anywhere and crack it open while you're waiting for soccer practice to start or for your friend to meet you for a study date or waiting for the library to open.* *ASAP Chemistry* is the perfect study guide for students who need high-level review in addition to their regular review and also for students who perhaps need to cram pre-exam. Whatever you need it for, you'll find no judgment here!

---

*Because you camp out in front of the library like they are selling concert tickets in there, right? Only kidding.

Who is This Book For?

This book is for YOU! No matter what kind of student you are, this book is the right one for you. How do you know what kind of student you are? Follow this handy chart to find out!

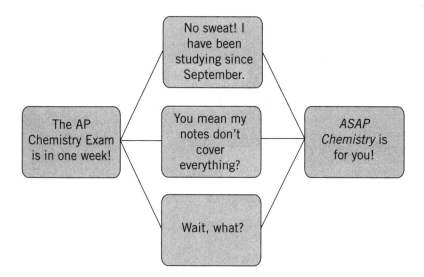

As you can see, this book is meant for every kind of student. Our quick lessons let you focus on the topics you must know, you should know, and you don't need to know—that way, even if the test is tomorrow (!), you can get a little extra study time in, and only learn the material you need.

How Do I Use This Book?

This book is your study tool, so feel free to customize it in whatever way makes the most sense to you, given your available time to prepare. Here are some suggestions:

Target Practice

If you know what topics give you the most trouble, hone in on those chapters or sections.

ASK Away

Answer all of the ASK questions *first*. This will help you to identify any additional tough spots that may need special attention.

Three-Pass System

Start at the very beginning!* Read the book several times from cover to cover, focusing selectively on the MUST content for your first pass, the SHOULD content for your second pass, and finally, the DO NOT NEED TO KNOW content.

 *It's a very good place to start.

Why Are There Icons?

Your standard AP course is designed to be equivalent to a college-level class, and as such, the amount of material that's covered may seem overwhelming. It's certainly admirable to want to learn everything—these are, after all, fascinating subjects. But every student's course load, to say nothing of his or her life, is different, and there isn't always time to memorize every last fact.

To that end, *ASAP Chemistry* doesn't just distill the key information into bite-sized chunks and memorable tables and figures. This book also breaks down the material into three major types of content:

🛑 This symbol calls out a section that has MUST KNOW information. This is the core content that is either the most likely to appear in some format on the test or is foundational knowledge that's needed to make sense of other highly tested topics.

💬 This symbol refers to SHOULD KNOW material. This is either content that has been tested in some form before (but not as frequently) or which will help you to deepen your understanding of the surrounding topics. If you're pressed for time, you might just want to skim it, and read only those sections that you feel particularly unfamiliar with.

🚫 This symbol indicates DO NOT NEED TO KNOW material, so make note of if, but don't worry about it! When the AP Chemistry course and exam were redesigned in 2013, some topics were removed from the curriculum. Some teachers or other prep books might not have been updated to match the new course and exam descriptions, so we are calling out any topics that you don't need to know (for the exam) with this icon. You may want to know certain topics or facts for your own chemistry edification and that's great, but if you're under a time crunch, steer clear of these items and don't spend time on them.

Introduction

As you work through the book, you'll also notice a few other types of icons.

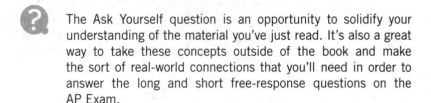

The Ask Yourself question is an opportunity to solidify your understanding of the material you've just read. It's also a great way to take these concepts outside of the book and make the sort of real-world connections that you'll need in order to answer the long and short free-response questions on the AP Exam.

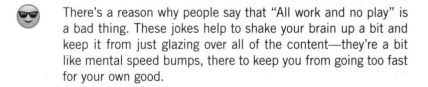

The Remember symbol indicates certain facts that you should keep in mind as you're going through the different sections.

There's a reason why people say that "All work and no play" is a bad thing. These jokes help to shake your brain up a bit and keep it from just glazing over all of the content—they're a bit like mental speed bumps, there to keep you from going too fast for your own good.

There's a lot to think about in this book, and when you see this guy, know that the information that follows is always good to have on hand. You'll rock it in trivia, if no place else.

Where Can I Find Other Resources?

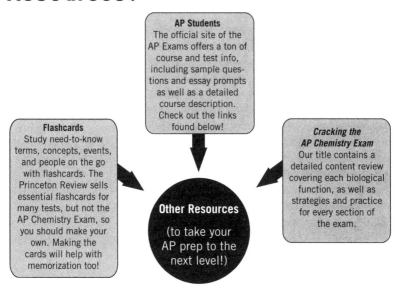

Useful Links

- AP Chemistry Homepage: https://apstudent.collegeboard.org/apcourse/ap-chemistry
- Your Student Tools: www.PrincetonReview.com/cracking
  See the "Get More (Free) Content" page for step-by-step instructions for registering your book and accessing more materials to boost your test prep.

CHAPTER 1
Atomic Structure

The chemical elements are fundamental building materials of matter, and all matter can be understood in terms of arrangements of atoms. In fact, atoms are the building blocks of all matter, and atoms retain their identity in chemical reactions. Understanding the structure of atoms and how atoms of different elements and isotopes differ from one another is important to your success on the AP Chemistry Exam!

Periodic Table 🛑

Let's start with a very fundamental part of chemistry—the periodic table. This section should ring lots of bells. The most important tool you will use on the AP Chemistry Exam is the Periodic Table of the Elements, which gives you important information about each element.

PERIODIC TABLE OF THE ELEMENTS

1 1 H 1.008	2											13	14	15	16	17	18 2 He 4.00
3 Li 6.94	4 Be 9.01											5 B 10.81	6 C 12.01	7 N 14.01	8 O 16.00	9 F 19.00	10 Ne 20.18
11 Na 22.99	12 Mg 24.30	3	4	5	6	7	8	9	10	11	12	13 Al 26.98	14 Si 28.09	15 P 30.97	16 S 32.06	17 Cl 35.45	18 Ar 39.95
19 K 39.10	20 Ca 40.08	21 Sc 44.96	22 Ti 47.90	23 V 50.94	24 Cr 52.00	25 Mn 54.94	26 Fe 55.85	27 Co 58.93	28 Ni 58.69	29 Cu 63.55	30 Zn 65.39	31 Ga 69.72	32 Ge 72.59	33 As 74.92	34 Se 78.96	35 Br 79.90	36 Kr 83.80
37 Rb 85.47	38 Sr 87.62	39 Y 88.91	40 Zr 91.22	41 Nb 92.91	42 Mo 95.94	43 Tc (98)	44 Ru 101.1	45 Rh 102.91	46 Pd 106.42	47 Ag 107.87	48 Cd 112.41	49 In 114.82	50 Sn 118.71	51 Sb 121.75	52 Te 127.60	53 I 126.91	54 Xe 131.29
55 Cs 132.91	56 Ba 137.33	57 *La 138.91	72 Hf 178.49	73 Ta 180.95	74 W 183.85	75 Re 186.21	76 Os 190.2	77 Ir 192.2	78 Pt 195.08	79 Au 196.97	80 Hg 200.59	81 Tl 204.38	82 Pb 207.2	83 Bi 208.98	84 Po (209)	85 At (210)	86 Rn (222)
87 Fr (223)	88 Ra 226.02	89 †Ac 227.03	104 Rf (261)	105 Db (262)	106 Sg (266)	107 Bh (264)	108 Hs (277)	109 Mt (268)	110 Ds (271)	111 Rg (272)							

	58 Ce 140.12	59 Pr 140.91	60 Nd 144.24	61 Pm (145)	62 Sm 150.4	63 Eu 151.97	64 Gd 157.25	65 Tb 158.93	66 Dy 162.50	67 Ho 164.93	68 Er 167.26	69 Tm 169.93	70 Yb 173.04	71 Lu 174.97
*Lanthanide Series														
†Actinide Series	90 Th 232.04	91 Pa 231.04	92 U 238.03	93 X (237)	94 Pu (244)	95 Am (243)	96 Cm (247)	97 Bk (247)	98 Cf (251)	99 Es (252)	100 Fm (257)	101 Md (258)	102 No (259)	103 Lr (262)

The horizontal rows of the periodic table are called **periods**. The vertical columns of the periodic table are called **groups**. Some important groups to know for the exam are the following:

- Group IA/1 – Alkali Metals
- Group IIA/2 – Alkaline Earth Metals
- Group B/3-12 – Transition Metals
- Group VIIA/17 – Halogens
- Group VIIIA/18 – Noble Gases

In addition, the two rows offset beneath the table are alternatively called the *lanthanides* and *actinides*, the rare Earth elements, or the inner transition metals.

Analysis of an Element

1. This is the **symbol** for the element; carbon, in this case. On the test, the symbol for an element is used interchangeably with the name of the element.
2. This is the **atomic number** of the element. The atomic number is the same as the number of protons in the nucleus of an element; it is also the same as the number of electrons surrounding the nucleus of an element when it is neutrally charged.
3. This number represents the average **atomic mass** of a single atom of carbon, measured in atomic mass units (amus). It also represents the average mass for a mole of carbon atoms, measured in grams. Thus, one mole of carbon atoms has a mass of 12.01 g. This is called the molar mass of the element.

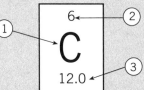

Atomic Identity

So, what does an atom look like? At the center of an atom is the nucleus, which is composed of protons and neutrons, and orbiting around the nucleus are electrons. The positively-charged nucleus is always pulling at the negatively-charged electrons around it, and the electrons have potential energy that increases with their distance from the nucleus.

Consider an atom of Helium, shown on the next page. The number of protons contained in its nucleus determines the identity of an atom; helium has two protons and is the 2nd element on the period table. The nucleus of an atom also contains neutrons. The mass number of an atom is the sum of its neutrons and protons; helium, with a mass number of 4, has a total of two neutrons and two protons. Electrons have significantly less mass than protons or neutrons and do not contribute to an element's mass.

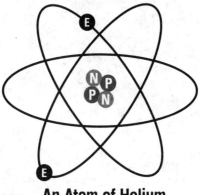

An Atom of Helium

Atoms of an element with different numbers of neutrons are called isotopes; for instance, carbon-12, which contains 6 protons and 6 neutrons, and carbon-14, which contains 6 protons and 8 neutrons, are isotopes of carbon. The **molar mass** given on the periodic table is the average of the mass numbers of all known isotopes weighted by their percent abundance.

Electron Configurations

Max Planck figured out that electromagnetic energy is **quantized**, which means that electrons can exist only at specific energy levels that are separated by specific intervals. That is, for a given frequency of radiation (or light), all possible energies are multiples of a certain unit of energy, called a **quantum** (mathematically, that's $E = h\nu$). So, energy changes do not occur smoothly but rather in small, specific steps.

Principal Energy Levels

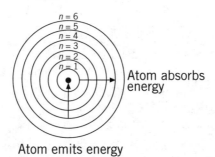

Neils Bohr took the quantum theory and used it to predict that electrons orbit the nucleus at specific, fixed radii, like planets orbiting the Sun. In the **Bohr Model**, each energy level is represented by a row on the periodic table. There are currently seven known energy levels, or **shells**, that correspond with $n = 1$ to $n = 7$. The closer an energy level is to an atom, the less energy electrons on that level have. The Bohr model is limited to two dimensions..

Conversely, the **quantum mechanical model** is a mathematical model that rejects the idea that electrons exist in only orbits around the nucleus. Instead, the quantum mechanical model describes the likelihood of finding an electron at different locations around the atom's nucleus. Unfortunately, watching reruns of the bonkers, hip, late 1980s/early 1990s TV show QUANTUM LEAP is not considered AP Chemistry Exam preparation.

Orbitals and Subshells

Negatively charged electrons are the smallest of the three subatomic particles that zoom around the nucleus in elliptical patterns called **orbitals**. Orbitals have associated energy levels, and are classified, in order from lowest to highest energy, as *s*, *p*, *d*, and *f* **subshells**. The maximum number of electrons a subshell can hold is as follows:

Subshell	Maximum Electrons
s	2
p	6
d	10
f	14

 There's nothing BOHRING about the BOHR Model, am I right?

Atomic Structure

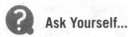

Ask Yourself...

What is the electron configuration for sulfur?

Sulfur has 16 electrons. The first two go into energy level 1 subshell s; this is represented by $1s^2$. The next two go into energy level 2 subshell s—$2s^2$. Six more fill energy level p ($2p^6$), then two more go into 3s ($3s^2$), and the final four enter into 3p ($3p^4$). So the final configuration is $1s^2 2s^2 2p^6 3s^2 3p^4$.

The periodic table is designed so that each area is exactly the length of one particular subshell.

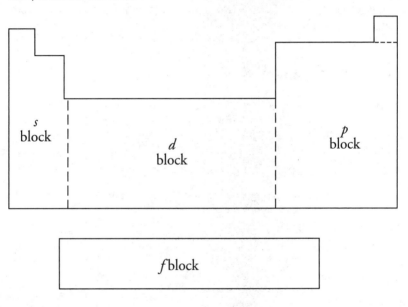

As you can see, the first two groups (plus helium) are in what is called the s-block. The groups on the right of the table are in the p-block, while transition metals make up the d-block. The inner transition metals below the table inhabit the f-block. The complete description of the energy level and subshell that each electron in an element inhabits is called its **electron configuration**.

 On this page we discussed four different blocks, and yet Jenny isn't from any of them!

Electron Configuration Principles

- **Aufbau Principle**: when building up the electron configuration of an atom, electrons are placed in orbitals, subshells, and shells in order of increasing energy.
- **Pauli Exclusion Principle**: two electrons which share an orbital cannot have the same spin. One electron must spin clockwise, and the other must spin counterclockwise.
- **Hund's Rule**: when an electron is added to a subshell, it will always occupy an empty orbital if one is available. Electrons always occupy orbitals singly if possible and pair up only if no empty orbitals are available. The given example shows how the $2p$ subshell fills when moving from boron to neon.
- **Transition Metals**: When entering the d-block, you should drop down one principal energy level. So, chromium's valence energy level would read $4s^2 3d^4$, not $4s^2 4d^4$.

	1s	2s	2p
Boron	⇅	⇅	↑
Carbon	⇅	⇅	↑ ↑
Nitrogen	⇅	⇅	↑ ↑ ↑
Oxygen	⇅	⇅	⇅ ↑ ↑
Fluorine	⇅	⇅	⇅ ⇅ ↑
Neon	⇅	⇅	⇅ ⇅ ⇅

Ion Formation

One of the great rules of chemistry is that the most stable configurations from an energy standpoint are those in which the outermost energy level is full. For anything in the s or p blocks, that means achieving a state in which there are eight electrons in the outermost shell (2 in the s subshell and 6 in the p subshell).

Atomic Structure

Elements that are close to a full energy level, such as the halogens or those in the oxygen group, tend to gain electrons to achieve a stable configuration. An **ion** is an atom that has either gained or lost electrons, while the number of protons and neutrons remains constant. Halogens need only gain one electron to achieve a stable configuration, and as such, typically form ions with a charge of negative one. Any particle with more electrons than protons is called an **anion**, or a negatively-charged ion.

On the other end of the table, the alkali metals can most easily achieve a full valence shell by losing a single electron, rather than by gaining seven. So, they will form positively-charged ions called **cations**, which have more protons than electrons. Alkali metals typically have a charge of +1, alkaline Earth metals of +2, and so forth. The table below gives a good overview of the common ionic charges for various groups:

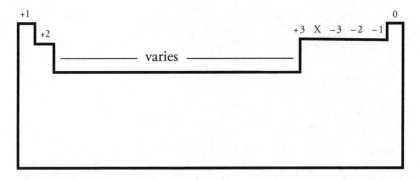

In general, transition metals can form ions of varying charges. All transition metals lose electrons to form cations, but how many electrons they lose will vary depending on the compound they are in. There are a few transition metals that only form ions with one possible charge. Two important examples are zinc, which always forms ions with a charge of +2, and silver, which always forms ions with a charge of +1.

There's also a slightly different rule as to how transition metals lose electrons when forming those cations. Transition metals, when losing electrons, will lose their higher-level *s* electrons before losing any of the lower-level *d* electrons.

Coulomb's Law 🛇

The amount of energy that an electron has depends on its distance from the nucleus of the atom; this can be calculated using **Coulomb's Law**. For the AP Chemistry Exam, you should be able to qualitatively apply Coulomb's Law given the charge of the nucleus.

Coulomb's Law

$$\mathcal{E} = \frac{k(+q)(-q)}{r^2}$$

$\mathcal{E}$ = energy
k = Coulomb's constant
$+q$ = magnitude of the positive charge (nucleus)
$-q$ = magnitude of the negative charge (electron)
r = distance between the charges

Electron Potential Energy

So, how does this relate to an electron's potential energy? Well, the greater the charge of the nucleus, the more energy an electron will have. Coulombic potential energy is considered to be 0 at a distance of infinity. For example, the Coulombic potential energy for a $1s$ electron is lower (more negative) than that of a $3s$ electron. The amount of energy required to remove a $1s$ electron, thereby bringing its Coulombic potential energy to zero, will thus be greater than the amount needed to remove a $3s$ electron. This removal energy is called the **binding energy** of the electron and is always a positive value.

Photoelectron Spectroscopy Concepts

If an atom is exposed to electromagnetic radiation at an energy level that exceeds the various binding energies of the electrons of that atom, the electrons can be ejected. The amount of energy necessary to do that is called the **ionization energy** for that electron. For the purposes of this exam, ionization energy and binding energy can be considered to be synonymous terms. When examining the spectra for electrons from a single atom or a small number of atoms, this energy is usually measured in **electronvolts**, eV (1 eV = 1.60×10^{-19} Joules). If moles of atoms are studied, the unit for binding energy is usually either kJ/mol or MJ/mol.

All energy of the incoming radiation must be conserved and any of that energy that does not go into breaking the electron free from the nucleus will be converted into kinetic energy for the ejected electron. The faster an ejected electron is going, the more kinetic energy it has. Electrons that were originally further away from the nucleus require less energy to eject, and thus will be moving faster. So, by examining the speed of the ejected electrons, we can determine how far they were from the nucleus of the atom in the first place.

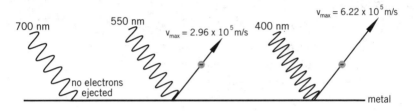

Incoming Radiation Energy = Binding Energy + Kinetic Energy
(of the ejected electron)

Usually, it takes electromagnetic radiation in either the visible or ultraviolet range to cause electron emission, while radiation in the infrared range is often used to study chemical bonds. Radiation in the microwave region is used to study the shape of molecules.

Interpreting PES

If the amount of ionization energy for all electrons ejected from a nucleus is charted, you get what is called a **photoelectron spectra** (PES). The PES for sulfur is as shown below.

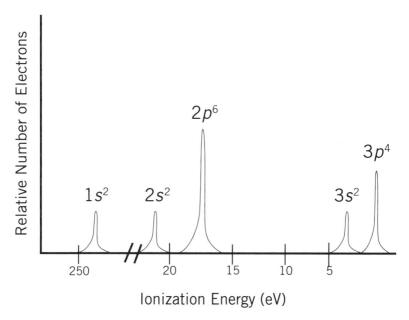

The y-axis describes the relative number of electrons that are ejected from a given energy level, and the x-axis shows the ionization energy of those electrons. Unlike most graphs, ionization energy decreases going from left to right in a PES.

Each section of peaks in the PES represents a different energy level. The number of peaks in a section shows us that not all electrons are the same distance from the nucleus. In this spectra, we can see the peak for the p-subshell in energy level 2 is three times higher than that of the s-subshell. The relative height of the peaks helps determine the number of electrons in that subshell. In the area for the third energy level, the p-subshell peak is only twice as tall as the s-subshell. This indicates there are only four electrons in the p-subshell of this particular atom.

Mass Spectrometry

The mass of various isotopes of an element can be determined by a technique called **mass spectrometry**. A mass spectrum of selenium looks like the graph below.

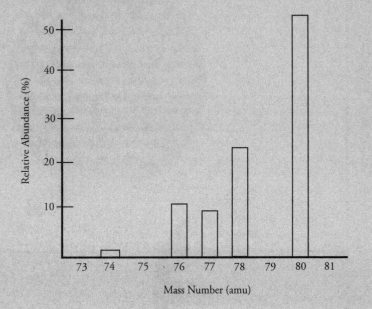

As you can see, the most abundant isotope of selenium has a mass of 80, but there are four other naturally occurring isotopes of selenium. The average atomic mass is the weighted average of all five isotopes of selenium shown on this spectra.

Studying Atoms

When atoms absorb energy in the form of **electromagnetic radiation**, electrons jump to higher energy levels. When electrons drop from higher to lower energy levels, atoms give off energy in the form of electromagnetic radiation. For a particular atom, the energy level changes of the electrons are always the same, so atoms can be identified by their emission and absorption spectra.

$$E = h\nu \text{ and } c = \nu\lambda$$

The relationship between the change in energy level of an electron and the electro-magnetic radiation absorbed or emitted is given below.

Energy and Electromagnetic Radiation

$$\varepsilon = h\nu = \frac{hc}{\lambda}$$

ε = energy change
h = Planck's constant, 6.63×10^{-34} joule·sec
ν = frequency of the radiation
λ = wavelength of the radiation
c = the speed of light, 3.00×10^{8} m/sec

The frequency and wavelength of electromagnetic radiation are inversely proportional. Combined with the energy and electromagnetic radiation equation, we can see that higher frequencies and shorter wavelengths lead to more energy.

Remember!

Unit Conversion!

Nanometers and meters are units of measuring length in the metric system. While nanometers are often used to measure distances in atoms and molecules, you may be asked to convert between nanometers (nm) and meters (m). Remember that the relationship between the two is as follows: 1 meter = 1,000,000,000 nanometers!

Let's take a look at an example:

a) Silicon has a ionization energy of 787 kJ/mol. How much energy, in Joules, is required to ionize a single silicon atom?

Solving this just requires a couple of simple unit conversions.

$$\frac{787 \text{ kJ}}{1 \text{ mol}} \times \frac{1 \text{ mol}}{6.02 \times 10^{23} \text{ atoms}} \times \frac{1000 \text{ J}}{1 \text{ kJ}} = 1.31 \times 10^{-18} \text{ J}$$

Atomic Structure

b) An atom of silicon is exposed to light with a wavelength of 250 nm. Will the atom be ionized?

First, we have to convert wavelength to frequency. To do so, wavelength needs to be in meters. 1 nm = 1×10^{-9} m, so 250 nm = 250×10^{-9} m. From here on out, it's just a math party!

$$c = v\lambda$$

3.0×10^8 m/s = $v(250 \times 10^{-9}$ m$)$
$v = 1.20 \times 10^{15}$ s^{-1}

$$E = hv$$

$E = (6.626 \times 10^{-34}$ Js$)(1.20 \times 10^{15}$ s$^{-1})$
$E = 7.95 \times 10^{-19}$ J

As 1.31×10^{-18} J $> 7.95 \times 10^{-19}$ J, the atom would not be ionized.

Frequency and Wavelength
$$c = \lambda v$$

c = speed of light (2.998×10^8 ms^{-1})
v = frequency of the radiation (in s^{-1})
λ = wavelength of the radiation (in m)

Periodic Trends 🛑

Now we're going to talk trends, but we don't mean trends like the latest rapper or dance craze (by the time we printed that in a book, everything would have changed anyway, though we'll gather our friends for a dance we call "Invisible Double Dutch" forever). We mean trends in the periodic table. Trends that essentially mean patterns.

You can make predictions about certain behavior patterns of an atom and its electrons based on the position of the atom in the periodic table. All the periodic trends can be understood in terms of three basic rules of electrons.

1. Electrons are attracted to the protons in the nucleus of an atom, and the closer an electron is to the nucleus, the more strongly it is attracted. The more protons in a nucleus, the more strongly an electron is attracted.

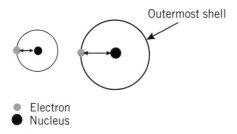

- Electron
- Nucleus

2. Electrons are repelled by other electrons in an atom. So, if other electrons are between a valence electron and the nucleus, the valence electron will be less attracted to the nucleus. That's called **shielding**.

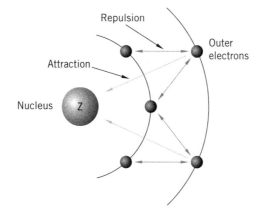

Atomic Structure

3. Complete shells are very stable. Atoms will add or subtract valence electrons to create complete shells if possible.

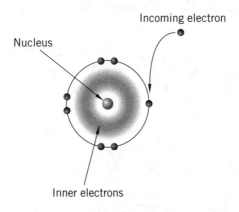

Isoelectronic Species

Let's break it down: isoelectronic species comes from the Latin 'iso' meaning same and 'electronic' which means charge. **Isoelectronic species** are atoms, molecules, or ions that have the same electronic structure and valence electrons. In general, isoelectronic chemical species exhibit similar chemical properties. The following chart shows isoelectronic ions and their electron configurations.

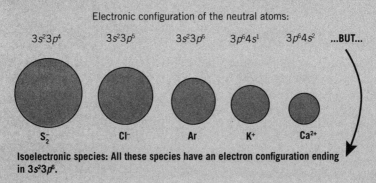

Isoelectronic species: All these species have an electron configuration ending in $3s^23p^6$.

Atomic Radius

The **atomic radius** is the approximate distance from the nucleus of an atom to its valence electrons. When moving from left to right across a period, atomic radius decreases; when moving down a group, atomic radius increases. Furthermore, cations have a smaller atomic radius than the atoms from which they are formed, and anions have a larger radius than the atoms from which they are formed.

Ionization Energy

The energy required to remove an electron from an atom is called the **first ionization energy.** Once an electron has been removed, the atom becomes a positively charged ion. The energy required to remove the next electron from the ion is called the second ionization energy, and so on. When an electron has been removed from an atom, electron-electron repulsion decreases, increasing the ionization energy.

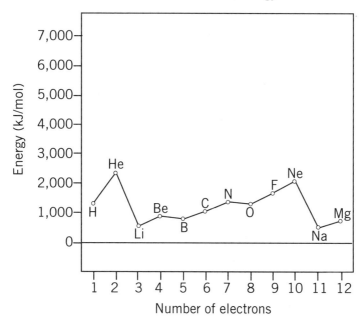

Additionally, a large jump in ionization energies accompanies a drop in principal energy levels. For instance, take a look at the first five ionization energies for aluminum:

Ionization	Energy (kJ mol)
1st	577
2nd	1,815
3rd	2,740
4th	11,600
5th	15,000

Notice the huge jump from the 3rd to 4th ionization? That occurs because after all three valence electrons from aluminum, all located on the third energy level, are removed, the next one to be removed has to come from the second energy level. That means the electron is much closer to the nucleus and requires a lot more energy to remove.

Electronegativity

Electronegativity refers to how strongly the nucleus of an atom attracts the electrons of other atoms in a bond. Electronegativity is affected by two factors: the size of an atom and the number of electrons in the atom's energy level. In general:

- Electronegativity increases when moving left to right across a period.
- Electronegativity decreases when moving down a group.

The primary exceptions to these trends are the electronegativity values for the three smallest noble gases. As helium, neon, and argon do not form bonds, they have zero electronegativity. The larger noble gases, however, can form bonds under certain conditions and do follow the general trends described above.

Decreasing Atomic Radius →

53 • H																	31 • He
167 • Li	112 • Be											87 • B	67 • C	56 • N	48 • O	42 • F	38 • Ne
190 • Na	145 • Mg											118 • Al	111 • Si	98 • P	88 • S	79 • Cl	71 • Ar
243 • K	194 • Ca	184 • Sc	176 • Ti	171 • V	166 • Cr	161 • Mn	156 • Fe	152 • Co	149 • Ni	145 • Cu	142 • Zn	136 • Ga	125 • Ge	114 • As	103 • Se	94 • Br	88 • Se
265 • Rb	219 • Sr	212 • Y	208 • Zr	198 • Nb	190 • Mo	183 • Tc	178 • Ru	173 • Rh	169 • Pd	165 • Ag	161 • Cd	156 • In	145 • Sn	133 • Sb	123 • Te	115 • I	108 • Xe
298 • Cs	253 • Ba	217 • La	208 • Hf	200 • Ta	193 • W	188 • Re	185 • Os	180 • Ir	177 • Pt	174 • Au	171 • Hg	156 • Tl	154 • Pb	143 • Bi	135 • Po	127 • At	120 • Rn

↑ Increasing Atomic Radius

↓ Increasing Ionization Energy

Calculated Atomic Radii (in Picometers)

Increasing Ionization Energy →

Increasing Electronegativity →

Bonding 💡

This kind of bonding isn't when you and your friends go to a cabin in the woods, eat popcorn, and share secrets—instead, this kind of bonding is all about sharing electrons (but now I'm hungry for popcorn).

Atoms engage in chemical reactions in order to reach a more stable, lower-energy state. This requires the transfer or sharing of electrons, a process that is called **bonding**.

Ionic

In an **ionic bond**, the cation gives up an electron, or electrons, to the anion. Electrostatic forces hold the two ions in an ionic bond together. In the diagram below, a sodium atom has given up its single valence electron to a chlorine atom, which has seven valence electrons and uses the electron to complete its outer shell (with eight). The two atoms are then held together by the positive and negative charges on the ions.

$$[\dot{Na}]^+ [:\ddot{Cl}:]^- \longrightarrow [Na]^+ [:\ddot{Cl}:]^-$$

Atomic Structure

The electrostatic attractions that hold together the ions in the NaCl lattice are very strong, and any substance held together by ionic bonds will usually be a solid at room temperature and have very high melting and boiling points. Two factors affect the melting points of ionic substances: ion charge and ion size. The primary factor is the charge on the ions; according to Coulomb's Law, a greater charge leads to greater bond energy, which is often called **lattice energy** in ionic bonds. If both compounds are made up of ions with equal charges, then the size of the ions must be considered; smaller ions will have greater Coulombic attraction.

In an ionic solid, each electron is localized around a particular atom, so electrons do not move around the lattice; this makes ionic solids poor conductors of electricity. Ionic liquids, however, do conduct electricity because the ions themselves are free to move about in the liquid phase, although the electrons are still localized around particular atoms. Salts are held together by ionic bonds.

Covalent

In a **covalent bond**, two atoms share electrons. Each atom counts the shared electrons as part of its valence shell. In this way, both atoms achieve complete outer shells. The number of covalent bonds an atom can form is the same as the number of unpaired electrons in its valence shell. The first covalent bond formed between two atoms is called a sigma (σ) bond. All single bonds are sigma bonds. If additional bonds between the two atoms are formed, they are called pi (π) bonds. The second bond in a double bond is a pi bond and the second and third bonds in a triple bond are also pi bonds. Double and triple bonds are stronger and shorter than single bonds, but they are not twice or triple the strength.

Summary of Multiple Bonds			
Bond type	Single	Double	Triple
Bond designation	One sigma (σ)	One sigma (σ) and one pi (π)	One sigma (σ) and two pi (π)
Bond order	One	Two	Three
Bond length	Longest	Intermediate	Shortest
Bond energy	Least	Intermediate	Greatest

In the diagram to the right, two atoms, each of which has one valence electrons, form a covalent bond. Each atom donates an electron to the bond, which is considered to be part of the valence shell of both atoms.

Usually, the more electronegative atom will exert a stronger pull on the electrons in the bond. This gives the molecule a **dipole**. That is, the side of the molecule where the electrons spend more time will be negative and the side of the molecule where the electrons spend less time will be positive.

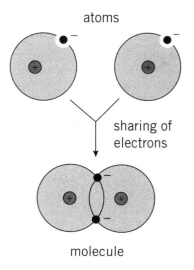

covalent bond

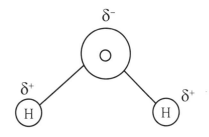

In the water molecule shown left, oxygen has a higher electronegativity than hydrogen and thus will have the electrons closer to it more often. This gives the oxygen a negative dipole and each hydrogen a positive dipole.

The **polarity** of a molecule is measured by the dipole moment. The more polar the molecule is, the larger the dipole moment is. You will not need to calculate the strength of a dipole, but you should be familiar with the unit with which that strength is quantified. That unit is called the **debye** (D).

 Remember!

Polarity Exception

Since carbon and hydrogen have very similar electronegativity values, C-H bonds are considered nonpolar. Because of this, the class of hydrocarbons is almost entirely non-polar.

Atomic Structure

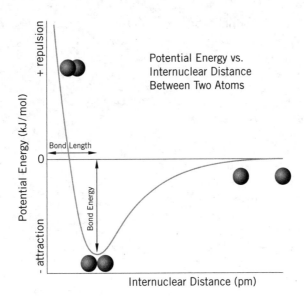

Covalent Bond Distance

Bond length, also referred to as **covalent bond distance,** is the average distance between the nuclei, or **internuclear distance,** of two bonded atoms in a molecule. It can be used to measure the strength of the bond, or how tightly two atoms are bonded. The bond length is the distance, d, at which the potential energy of the two atoms is at a minimum. The greater the internuclear distance, the lower the potential energy; i.e., when d is small, the stronger the bond is between two atoms, and the higher the potential energy. The graph to the right shows the relationship between internuclear distance and potential energy.

Network Covalent Bonds

In a network solid, atoms are held together in a lattice of covalent bonds. Network solids are very hard and have very high melting and boiling points. The electrons in a network solid are localized in covalent bonds, which makes them poor conductors of electricity. The most commonly seen network solids are compounds of carbon (such as diamond or graphite) and silicon (SiO_2—quartz).

Like athletes who dope in the Olympics to improve performance (a practice that is rare and shameful), doping in chemistry adds an impurity to a lattice to increase conductivity. **Doping** is a process in which an impurity is added to an existing lattice. In a normal silicon lattice, each individual silicon atom is bonded to four other silicon atoms. When some silicon atoms are replaced with elements that have only three valence electrons (such as boron or aluminum), the neighboring silicon atoms will lack one bond apiece. This missing bond (or "hole") creates a positive charge in the lattice, and the hole attracts other electrons to it, increasing conductivity. Those electrons leave behind holes when they move, creating a chain reaction in which the conductivity of the silicon increases. This type of doping is called **p-doping** for the positively charged holes.

If an element with five valence electrons (such as phosphorus or arsenic) is used to add impurities to a silicon lattice, there is an extra valence electron that is free to move around the lattice, causing an overall negative charge that increases the conductivity of the silicon. This type of doping is called **n-doping** due to the free-moving negatively charged electrons.

If an element with five valence electrons (such as phosphorus or arsenic) is used to add impurities to a silicon lattice, there is an extra valence electron that is free to move around the lattice, causing an overall negative charge that increases the conductivity of the silicon. This type of doping is called **n-doping** due to the free-moving negatively charged electrons.

Metallic

The positively-charged core of a metal is generally stationary, while the valence electrons are very mobile. The delocalized structure of a metal also explains why metals are both malleable and ductile. Metals can also bond with each other to form **alloys**. This typically occurs when two metals are melted into their liquid phases, and are then poured together before cooling and creating the alloy. In an **interstitial alloy**, metal atoms with two vastly different radii combine; e.g., the much smaller carbon atoms occupy the interstices of iron atoms to create steel. A **substitutional alloy** forms between atoms of similar radii.

Interstitial Alloy

Substitutional Alloy

CHAPTER 2
Covalent Bonding and Intermolecular Forces

In the first chapter, you reviewed that elements are the fundamental building blocks of matter, and that all matter can be understood in terms of its arrangement of atoms. Since the chemical and physical properties of matter can be explained by the arrangement of atoms, ions, or molecules and the forces between their neighboring particles, in this chapter, we will review the basics of covalent bonding and the forces of attractions between molecules.

Sharing is Caring 🔴

In order to achieve a stable noble gas configuration, nonmetal atoms often combine by sharing electron pairs. Remember, atoms will lose, gain, or share electrons in order to have a full valence shell of eight electrons called the **Octet Rule**. A covalent bond is a bond in which, one or up to three electron pairs is shared by the nuclei of two atoms.

For example, a fluorine atom has seven valence electrons and is unstable. Since atoms containing eight electrons in their valence shell are stable, one way for a fluorine atom to become stable is for two fluorine atoms to combine. Each atom contributes a valence electron to a shared electron pair and thus two fluorine atoms form a single covalent bond. The number of covalent bonds each atom can form is determined by the number of unpaired electrons in the atoms valence shell.

A pair of fluorine atoms forms an F_2 molecule in which each atom has a total of eight valence electrons by sharing a pair of electrons.

$$:\!\ddot{F}\!: \;+\; :\!\ddot{F}\!: \;\longrightarrow\; :\!\ddot{F}\!\!:\!\!\ddot{F}\!:$$

electron pair shared between the two fluorine atoms

Lewis Structures 🔴

Lewis structures (also called **electron dot diagrams**) are used to depict both shared and lone pairs of electrons in molecules. A pair of electrons that are shared between two atoms is called a bonded pair. A pair of electrons that are not shared between atoms is called a **lone pair** of electrons. The shared electron pairs in covalent bonds are indicated by pairs of dots between two atoms or lines for simplicity and the lone pairs are indicated by dots around one atom.

Summary of Steps for Drawing Lewis Structures

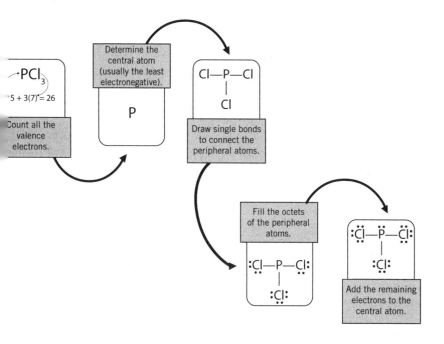

Examine the central atom of the structure:

1. If the central atom has a full octet, the Lewis structure is complete.
2. If the central atom has fewer than eight electrons, remove an electron pair from one of the peripheral atoms and place a line between the peripheral and central atom so that the electron pair is shared between the two atoms. You may have to remove additional electron pairs from other peripheral atoms to ensure all of the atoms have a complete octet.
3. If the central atom has more than eight electrons, that is okay as long as the number of electrons on the central atom does not exceed twelve or the number of valence electrons you calculated in Step 1. Note that only elements from period 3 or lower on the periodic table can have more than eight electrons.

Bond Order & Type

Each covalent bond has a fixed and characteristic distance between the two bonded nuclei called the bond length. Further, covalent bonds between different atoms have characteristic bond energies, which are inversely proportional to the bond length. To break it down, think about it like this. You have many relationships of varying degrees of closeness. You are closer to those people with whom you have a strong relationship. In a covalent bond, the bond strength is greater if the distance between the atoms sharing the electrons is shorter. Now I know you may feel close to friends that live far away, but remember atoms don't have feelings! The bond energy is the energy required to break such a bond, and also the energy liberated when such a bond is formed.

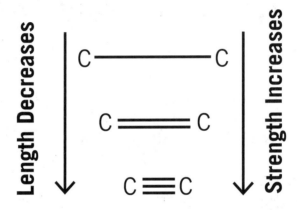

Summary of Multiple Bonds			
Bond type	**Single**	**Double**	**Triple**
Bond designation	One sigma (σ)	One sigma (σ) and one pi (π)	One sigma (σ) and two pi (π)
Bond order	One	Two	Three
Bond length	Longest	Intermediate	Shortest
Bond energy	Least	Intermediate	Greatest

Resonance Structures Involve the Shifting of Electrons

When it is possible to draw more than one acceptable Lewis structure for a given molecule, where all the connections are the same in both structures but the location of the electrons is different, we call these structures **resonance structures**. These structures are hypothetical and only exist in theory. That is, the molecule does not actually go back and forth between these configurations; rather, the true structure is an approximate intermediate of all the structures. One important thing to know about resonance structures is that the bond lengths and strengths of the bonds in resonance are equivalent. For example, in the resonance structures for ozone (O_3) below, the drawing makes it looks as if both structures contain a single O-O bond and a double O-O, but the reality is that both O-O bonds have both partial single/double character. In simple terms, both O-O bonds are stronger/shorter than a single O-O bond, but weaker/longer than a double O-O bond, an intermediate between the two.

Resonance Structure of Ozone, O_3

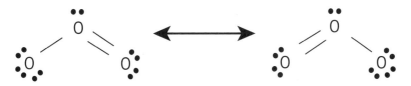

Lewis Structure of O_3	Expected	Actual
1 double bond and 1 single bond	1 short bond and 1 long bond	bonds are of equal length (128 pm)

Covalent Bonding and Intermolecular Forces

Isomers Involve the Rearrangment of Atoms

When a molecule is connected together with bonds, sometimes there is more than one way to arrange the atoms. Isomers are molecules with the same chemical formula, but the atoms are bonded together in a different spatial arrangement. Glucose, galactose, and fructose all have the same molecular formula, $C_6H_{12}O_6$.

Carbohydrate Isomers

Glucose:
$$\begin{array}{c} H-C=O \\ | \\ H-C-OH \\ | \\ HO-C-H \\ | \\ H-C-OH \\ | \\ H-C-OH \\ | \\ CH_2OH \end{array}$$

Galactose:
$$\begin{array}{c} H-C=O \\ | \\ H-C-OH \\ | \\ HO-C-H \\ | \\ HO-C-H \\ | \\ H-C-OH \\ | \\ CH_2OH \end{array}$$

Fructose:
$$\begin{array}{c} CH_2OH \\ | \\ C=O \\ | \\ HO-C-H \\ | \\ H-C-OH \\ | \\ H-C-OH \\ | \\ CH_2OH \end{array}$$

Advanced Lewis Strucutures

Formal Charge

Formal charge is an accounting procedure, but the currency is electrons. It enables chemists to determine the location of charge within a molecule. When atoms bond, they each bring some number of valence electrons into the molecule but keeping that same number of electrons close to it is important in order to balance the charge of the positive protons in its nucleus. So, if an atom ends up with more or fewer electrons around it, it can end up with a negative or positive formal charge, respectively. If a molecule has formal charges, they all must add up to the total charge on a molecule, which is zero. Lewis structures for a molecular cation or an anion must contain at least one atom with a formal charge, and sum of the formal charges on the individual atoms must sum to the charge of the ion. The formula for calculating formal charge on an atom is shown below.

VE = # of valence electrons
CB = # covalent bonds
LE = # non-bonding electrons

Formal Charge = VE − CB − LE

$[:C \equiv N:]^-$	**Example Molecule: Cyanide Ion**
Valence e^-: 4 Number of covalent bonds: 3 Number of lone electrons: 2	**Formal Charge on Carbon** $4 − (3) − (2) = −1$
Valence e^-: 5 Number of covalent bonds: 3 Number of lone electrons: 2	**Formal Charge on Nitrogen** $5 − (3) − (2) = 0$
The overall charge on the ion must be the sum of the formal charges on the atom.	**Overall Formal Charge** $−1 + 0 = −1$

Using Formal Charge to Determine the Best Lewis Structure

The best Lewis structure...

- Is the one with the fewest charges
- Puts a negative charge on the most electronegative atom

Violations of the Octet Rule

Following the Octet Rule leads to the most accurate representations of stable molecules and therefore we want to abide by the Octet Rule, if possible.

The Octet Rule is violated in three scenarios:

Odd valence

When there are an odd number of valence electrons, i.e, nitrogen containing oxides (Note: odd numbers of valence electrons won't be covered on the AP Chem Exam, so don't sweat this too much.)

Incomplete Octets

When there are fewer than eight valence electrons, i.e., H and B

Expanded Octet

When there are too many valence electrons, i.e., when the central atom is in the 3rd period or below and expanding its octet eliminates some formal charges

Covalent Bonding and Intermolecular Forces

⊘ Odd number of valence electrons: Free Radicals

Having an odd number of electrons is a violation of the Octet Rule because the rule requires that each atom has eight valence electrons surrounding it. Molecules containing an odd number of electrons have unpaired electrons and are known as **free radicals**. While many of these molecules are highly reactive and unstable, some are stable for days, months of even years. The most common free radicals that you will encounter are nitrogen containing oxides. For example, nitrogen monoxide contains 11 valence electrons (5 from nitrogen and 6 from oxygen). No matter how electrons are shared between the nitrogen and oxygen atom, there is no way for nitrogen to have an octet. It will have seven electrons instead.

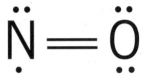

 Ask Yourself...

Why is nitrogen the recipient of the odd electron in the above Lewis structure for nitrogen monoxide?

Incomplete Octets and Expanded Octets

There are a few elements that form stable compounds without obtaining a valence of eight electrons. When this occurs the molecule contains an **incomplete octet**. First, H and He follow a "duet rule" because they only contain one energy level and therefore can only hold two electrons in their valence shell. Remember that H can achieve the noble gas confirguration of He by sharing an electron with only one bonding partner. The other notable exception is B, which can form stable compounds with only six valence electrons, such as BF_3. BF_3 is stable, but it will form BF_4^- when possible (and this ion, in which B has an octet, has shown up on the exam).

More common than incomplete octets are **expanded octets** where the central atom in a Lewis structure has more than eight electrons in its valence shell. Molecules with expanded octets involve highly electronegative terminal atoms, and a nonmetal central atom that is found in the third period or beyond. Elements in the third period have $3s$ and $3p$ orbitals. The $3d$ orbital is empty and therefore it can accommodate extra electrons. Since elements in the second period only have $2s$ and $2p$ orbitals, they cannot have more than eight electrons in their valence orbitals.

One of the situations where expanded octet structures are treated as more favorable than Lewis structures that follow the Octet Rule is when the formal charges in the expanded octet structure are smaller in magnitude than in a structure that adheres to the Octet Rule, or when there are less formal charges in the expanded octet than in a structure that adheres to the Octet Rule.

For example, below are two potential Lewis diagrams for ClO_3^-. Both have the correct number of valence electrons (32), and both show all atoms with a complete octet. However, in Structure A, all three oxygen atoms would have a formal charge of –1, and the chlorine would have a formal charge of +2. In Structure B, only one oxygen atoms have a formal charge of –1, and the chlorine has no formal charge. Structure B is thus the more likely structure.

Structure A   **Structure B**

Note that you DO NOT need to consider expanding octets when drawing Lewis structures just to minimize formal charge. However, if asked to choose between two structures, then you need to consider formal charge.

Molecular Geometry

Unfortunately, Lewis Structures only give us a two-dimensional image of a molecule. To fully grasp the physical interactions and chemical reactivity that a molecule will engage in with other molecules, we need to represent the molecule in three dimensions. Chemists use a simple model called "**Valence Shell Electron Pair Repulsion Theory**," or **VSEPR** that allows us to predict the 3-D shape of a molecule using its Lewis Structure. VSEPR is based on the idea that electron pairs surrounding the central atom want to be as far apart from each other as possible. This makes sense because negatively charged particles repel each other. This means that electrons surrounding the central atom will distribute themselves around the central atom with angles between them as far as possible, which gives rise to different molecular shapes. Importantly, the degree of repulsion between areas of electrons (electron domains) actually depends on whether the electrons are bonding or non-bonding electrons (lone pairs). Lone pairs by far have significantly greater repulsion amongst other electrons and will alter the bond angle and molecular shape to the greatest extent.

The order of electron-pair repulsions from greatest to least repulsion is

lone pair – lone pair > lone pair – bonding pair > bonding pair – bonding pair

---

Steps for Using VSEPR to Determine Molecular Shape

1. Draw the correct Lewis Structure of the molecule or ion.
2. Count the regions of electron domains around the central atom. A single, double, or triple bond as well as a lone pair all count as one electron density.
3. Use the number of lone pairs to determine the effects on the electron repulsion and resulting molecular shape.

Summary of VSEPR shapes

number of electron domains	0 lone pairs	1 lone pair	2 lone pairs	3 lone pairs	4 lone pairs
2	Linear 180° Angle				
3	Trigonal planar 120° Angle	Bent < 120° Angle			
4	Tetrahedral 109.5° Angle	Trigonal pyramidal < 109.5	Bent << 109.5		
5	Trigonal bipyramidal	Seesaw	T-shaped	Linear	
6	Octahedral	Square pyramidal	Square planar		

Note that when on a central atom, lone pairs exert a greater repulsive force than a bonded pair, which explains the decreased bond angles.

Below are a few molecules along with their corresponding VSEPR shapes.

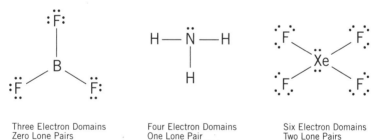

Three Electron Domains
Zero Lone Pairs
Trigonal Planar

Four Electron Domains
One Lone Pair
Trigonal Pyramidal

Six Electron Domains
Two Lone Pairs
Square Planar

Covalent Bonding and Intermolecular Forces 37

Polarity

Polarity arises from differences in electronegativity between the bonded atoms. By definition, electrognegativity is the ability of an atom to attract a bonding pair of electrons to its own nucleus. However, the best way to remember it is that electrognegativity is greed for electrons! Electrons are the currency of chemistry. More than any other factor, the electronegativity of the atom determines the chemical personality of the atom (how it bonds, the polarity of the bonds, the overall polarity of the molecule, and how the molecules interact with others.) Yeah, it is super important!

Electronegativity values are used as a guide to indicate whether the electrons in a bond are equally shared or unequally shared between two atoms. When electrons are shared equally, the bond is nonpolar. When differences in electronegativity result in unequal sharing of electrons (basically when the two atoms are not the same), the bond is polar, and is said to have a separation of charge or a "dipole." The greater the electronegativity difference, the more polar the bond. You will never need to know the electronegativity values, but remember the trend from the previous chapter; electronegativity increases as you move up and and to the right of the periodic table.

Drawing Dipoles

One way to indicate the polarity of the bond is to draw the dipole. A dipole can be represented by drawing an arrow over the bond pointing to the more electronegative atom. A "+" is placed at the tail of the arrow to indicate the partial positive end of the bond. Chemists may also use the symbol "δ+" to indicate the "electron deficient" atom and "δ-" to indicate which atom is "electron rich." For example, HCl is a diatomic molecule with a polar bond in which the bonding electrons spend more time around Cl than H.

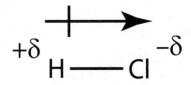

Symmetry Predicts Polarity

One way to determine if a molecule is polar or not is to examine its overall shape. Diatomic molecules with two identical atoms are nonpolar, but other shapes are symmetrical and therefore, also nonpolar. The symmetry in the carbon dioxide molecule shown below makes the molecule nonpolar even though it contains polar bonds. This is because the dipoles on the molecule are equal and opposite and therefore cancel each other out.

Symmetry has two components:

- the geometric arrangement of the peripheral atoms, and
- whether or not they are all the same.

Dipoles cancel if the molecule is symmetrical and all of the perpherial atoms are the same. Carbon dioxide is nonpolar because its equal and opposite dipoles cancel.

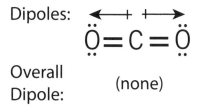

Dipoles: ←—+ +—→

$\ddot{O} = C = \ddot{O}$

Overall Dipole: (none)

Steps for Determining Polarity

1. Is the molecule a hydrocarbon (contains only C and H)?
   a. Yes. It is nonpolar. (Note: The electronegativity difference between C and H is neglible.)
   b. No. Draw the molecule's Lewis Structure.
2. Use VSEPR to determine the shape of the molecule.
3. Identify whether the shape is symmetrical (See the next page).
4. If all the peripheral atoms on the molecule are the same and the shape is symmetrical; then the molecule is nonpolar.

Molecular Symmetry Helps to Determine Polarity

Symmetrical Shapes

All of the symmetrical shapes (linear, trigonal planar, tetrahedral, trigonal bipyramidal, and octahedral) yield non-polar molecules as long as all of the peripheral atoms are identical. These shapes are associated with molecules whose central atoms contain no lone pairs.

Asymmetrical Shapes

All of the asymmetrical shapes (bent, trigonal pyramidal, seesaw, T-shaped, square pyramidal). Theses shapes are associated with molecules whose central atoms contain one or more lone pairs.

Exception

Square planar molecules have lone pairs on the central atom, but are still considered symmetrical.

Ask Yourself...

Did you know that polar molecules have a smell, whereas nonpolar molecules are undetectable to the human nose? What does that say about the polarity and the molecular shape of the mucous membrane in your nose?

Hybridization

Hybridization occurs when covalent bonds are shared. Briefly, when bonds in an *s*-orbital mix with those in *p*-orbitrals, we call the bond that forms a hybrid orbital. All you need to know for the exam is how to determine hybridization, and to do that, you count the number of electron domains around the central atom.

Number of electron domains	Hybridization
2	*sp*
3	*sp²*
4	*sp³*

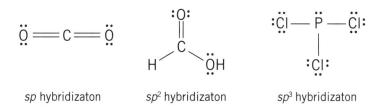

sp hybridizaton   *sp²* hybridizaton   *sp³* hybridizaton

🚫 Hybridization of expanded octets is NOT part of the AP curriculum.

Intermolecular Forces (IMFs)

Whereas **bond strength** describes strength of the chemical bonds within molecules, **intermolecular forces** describe the physical attraction between neighboring molecules. Here's an analogy to help you differentiate the two. You may be initially attracted to a "neighbor" solely based on physical appearance alone. (Love at first sight, right?) Well, the degree of physical attraction between molecules determines the physical properties of a substance and the energy required to change the physical state of a substance. In a solid, the particles are vibrating and have low energy, but because the particles are so strongly attracted, they are not free to move. There is less physical attraction between liquid particles, which also have more kinetic energy and therefore the ability to flow. Lastly, gas particles have the most kinetic energy and with little to no physical attraction, they have complete freedom to move. Note: the physical state of a substance is temperature dependent because temperature is directly proportional to the average kinetic energy of the particles.

So let's say that we have some particles in various physical states hanging out at a party. In the case of solids, the particles are so physically atrracted that they don't leave each others' sides all night. In the case of liquids, the particles stick close by, but since they are not as attracted, they mingle with one another, but still stick close by. On the other hand, in gases the particles are so not attracted, so they come and go readily without any physical attraction. Gas is cool as a cucumber, man—easy come, easy go.

Particulate Representations of IMFs

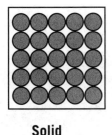 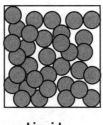 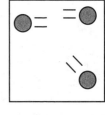

      Solid   **Liquid**   **Gas**

Generally, INTERmolecular forces (IMFs) are much weaker than INTRAmolecular forces (i.e., bonds). For example, it will take 41 kJ to vaporize 1 mole of water; whereas it will take 930 kJ to break all O-H covalent bonds in 1 mole of water.

IMFs are forces of physical attraction between neighboring molecules, not bonds within a single molecule.

The pairs of arrows within the water molecules below denote polar covalent bonds within the molecule which are intramolecular forces; whereas the single set of arrows between the two water molecules denotes the hydrogen "bonding" attraction that occurs between water molecules.

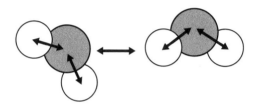

IMFs are forces of physical attraction BETWEEN interacting molecules. They are not bonds! They are drawn differently too! Solids lines represent covalent bonds, whereas dotted lines represent IMFs!

$$\underset{H}{\overset{\delta+}{}} \longrightarrow \underset{Cl}{\overset{\delta-}{}} \cdots\cdots\cdots \underset{H}{\overset{\delta+}{}} \longrightarrow \underset{Cl}{\overset{\delta-}{}}$$

------ Intermolecular force
——— Intramolecular force (bonds)

Covalent Bonding and Intermolecular Forces

Four Main Types of IMFS

1. Permanent dipoles

The positive end of a polar molecule is attracted to the negative end of its neighboring polar molecule. The permanent dipole adds to the attractive forces between the molecules raising the boiling and melting points relative to nonpolar molecules of similar size and shape.

2. Hydrogen "bonding"

Hydrogen "bonding" occurs when a hydrogen atom is bonded to an extremely electronegative element (N, O, or F). The hydrogen gains a very strong positive dipole, which is then attracted to the negative dipole on the N, O, or F in a different molecule.

Hydrogen bonding—IT IS NOT A BONDED HYDROGEN; it is an attraction between an unshielded hydrogen and an electronegative atom of its neighbor.

3. London Dispersion forces

In all chemical species, the electrons are constantly in motion. This motion creates fluctuations in the electron distribution in atoms and molecules that results in temporary dipoles. The region of the molecule with excess electron density has partial (−) charge, whereas the region with depleted electron density has partial (+) charge.

4. Dipole-induced dipole

This weak attraction occurs when a polar molecule induces a temporary dipole in an atom or in a nonpolar molecule with no dipole by disturbing the arrangement of electrons in that molecule.

Dipole-Dipole Interactions

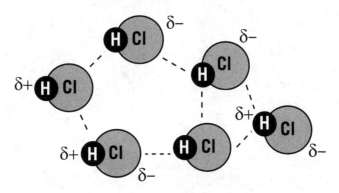

Water can form up to four hydrogen "bonds" with nearby water molecules.

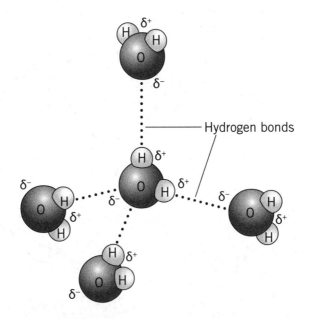

LDF strength increases as the number of electrons increase.

→ Weakest IMFs → Strongest IMFs

F-F   Br-Br   I-I

→ INCREASING # of electrons →

Liquid at RT   Solid at RT

As shown above, as the number of electrons increase in the halogens, so do the IMFs. Fluorine has the weakest IMFs and is a gas a room temperature. Bromine has stronger IMFs and is a liquid at room temperature. Finally, iodine has the strongest IMFs and is a solid at room temperature.

Remember!
Nonpolar molecules with a lot of electrons can have stronger IMFs than highly polar molecules with very few electrons.

Summary of IMFs

IMF Type	Basis of Attraction	Relative Strength
Hydrogen "Bonding"	Molecules having H bonded to F, O, or N attracted to neighboring molecules with dipoles	Strong
Dipole-Dipole	Two polar molecules attract	Moderate
London Dispersion Forces (LDFs)	Polarizable e^- clouds	Weak, but increases with increasing e^-
Dipole-induced dipole	A polar molecule and a nonpolar molecule	Weak

Covalent Bonding and Intermolecular Forces

Effects of IMFs on Physical Properties

Since the strength of the IMFs in a substance determines how well molecules stick together, this affects many of the physical properties of a substance.

Vapor Pressure

stronger the IMFs → lower vapor pressure

Vapor pressure arises from the fact that the molecules inside a liquid are in constant motion. There is always a small amount of gas that is found above all liquids. If the liquid molecules hit the surface of the liquid with enough kinetic energy, they can escape the intermolecular forces holding them to the other molecules and transition into the gas phase. The vapor pressure at given temperature is determined by the IMFs in the liquid. Weaker IMFs result in higher vapor pressure because the liquid molecules are not held together as strongly and are able to escape from the surface.

Melting & Boiling Point

stronger IMFs → higher melting and boiling points

If molecules stick together more, more energy will be required to break them apart. Some IMFs must be broken in order to melt a substance, and all IMFs must be overcome in order to vaporize a substance.

Surface Tension

Surface tension is the amount of energy required to increase the surface area of a liquid. Note the correlation between the surface tension of a liquid and the strength of the intermolecular forces: the stronger the IMFs, the higher the surface tension.

Capillary Action

Capillary action is the ability of a liquid to climb the walls of very narrow tubes due to the strong adhesive forces between the liquid and the tube. Capillary action only works if the adhesive forces outweigh cohesive forces between the liquid molecules.

How does water defy gravity? First, water forms strong adhesive forces with the glass of the narrow tube through hydrogen bonds with the oxygen atoms in glass (SiO_2). Second, if you use a narrow tube, the water molecules at the edges have fewer other water molecules to drag up the tube than in a large tube. Therefore, water can rise higher in a narrow tube than in a wider tube. In contrast, mercury will not rise up a narrow tube because its cohesive forces between atoms are stronger than its adhesive forces with the tube.

Ranking the Strength of IMFs

If you are asked to rank molecules in order of melting point, boiling point, vapor pressure, or surface tension, what they are actually asking is for you to rank them by strength of intermolecular forces (either increasing or decreasing). Here is a quick strategy for ranking:

1. Look for molecules that exhibit hydrogen bonding. They will have the strongest IMFs. Due to the presence of an –OH (hydroxyl) group or –NH_2 (amino) group, alcohols and amines can hydrogen bond. This leads to higher boiling points compared to nonpolar hydrocarbon chains of similar size.
2. Look for polar molecules (those with dipoles). These will have the next strongest IMFs.
3. If all molecules are nonpolar, then look for larger molecules. London dispersion forces are the weakest IMFs; however, larger molecules have more electrons and therefore have a more polarizable electron cloud and stronger IMFs than smaller molecules.

Covalent Bonding and Intermolecular Forces

Solubility

Two substances form a homogeneous mixture known as a **solution** when there is an attraction between the particles of the solute and solvent. When it comes to solubility, which is the ability of a solute to dissolve in a solvent, there is only one major rule: "Like dissolves like."* Polar solutes (such as sugars) or ionic solutes (such as ionic salts) dissolve in polar solvents such as water. Nonpolar solutes such as fats and oils will dissolve in nonpolar solvents such as hexane (C_6H_6).

Graphic Solubility of Compounds is Determined by "Like" Polarity

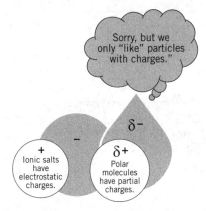

Ionic and polar molecules "like" each other and are miscible in one another.

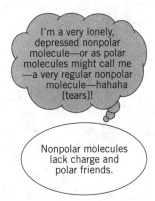

Nonpolar molecules lack charge and therefore do not mix with ionic and polar substances.

 *We might sound like your parents telling you to please say "like" less often, but it's undeniable that like dissolves like!

Even though like dissolves like, there is a limit to the amount of the solute that can be dissolved by a given amount of solvent. The term **solubilty** describes the amount of solute that will dissolve in a given solvent at a given temperature. Suppose we wish to make a solution of sodium chloride in water. If we add a few grams of water and stir, all of the NaCl particles will dissolve. Since more NaCl can dissolve, we call this an **unsaturated solution**. The solubility of NaCl is about 36 grams per 100 grams of water at 30°C, which means that after this quanity of NaCl has been added to 100 grams of water, no additional NaCl will dissolve and thus the solution is saturated. If we somehow manage to add more NaCl to the saturated solution and get it to dissolve (usually via heating or cooling), this is called a **supersaturated solution**.

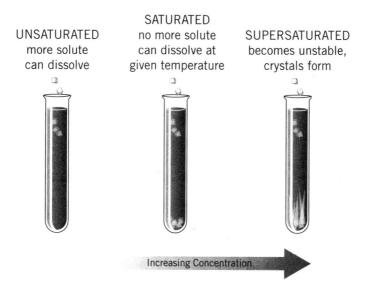

Covalent Bonding and Intermolecular Forces

Relationships between Solubility and Temperature

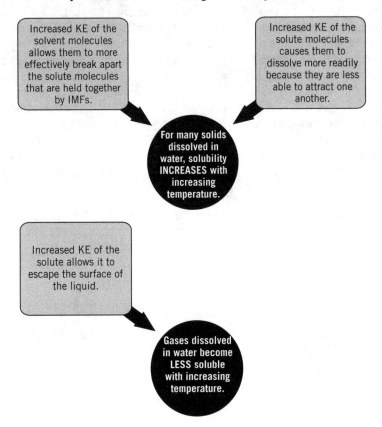

 Ask Yourself...

Why is increasing the temperature of the water a problem for the survival of aquatic life?

Crystalline vs. Amorphous Solids

In general, the particles of solid matter are held together by the strong attractive forces between them.

The components of a solid are usually arranged in two general ways:

- repeating arrays of a three-dimensional structure called a crystal lattice, thus forming a crystalline solid

or

- the aggregates of particles in no particular order called an amorphous solid.

Crystalline vs. Amporhous SiO_2

Crystalline SiO_2
(Quartz)

Amorphous SiO_2
(Glass)

· Si   • O

All you need to know about crystalline and amorphous properties is the general definition of the terms, so don't worry about anything else!

Covalent Bonding and Intermolecular Forces

Biological System Properties

Enzyme Catalysis

The cells in your body are tiny chemical factories performing hundreds of chemical reactions in order to carry out the functions necessary for life. These reactions are facilitated by **enzymes**, which are complex biological molecules that function as catalysts to biochemical reactions. While most enzymes are globular proteins, some enyzmes are made of ribonucleic acids and are called ribozymes. Like all cataylsts, enzymes do not change the position of equibrium, they increase the rate of reaction by reducing the amount of energy needed for the reactants to come together and react (**activation energy**) without being altered or consumed by the reaction.

Unlike other catalysts, enzymes contain a specific pocket for the reactant to fit into called an **active site**. Think of an assembly line in a factory—for efficiency, each worker does one very specific job over and over. The enzymes in your cells are just like that. The molecule on which the enzyme works, and whose reaction it speeds up, is called a **substrate**. When the substrate binds to the active site on the enzmye, structural changes in the enzyme result in the stabilization of the transiton state complex. This stabilization provides an alternate pathway which is lower in activation energy and therefore speeds up the reaction rate. Once the products leave the active site, the enzyme is ready to attach to a new substrate and repeat the process just like an assembly line worker!

Enzymes

Speed Up Reactions: Enzymes lower the activation energy, which increases the rate of a biochemical reaction.

Are Highly Specific: Enzymes are very specific catalysts and usually work to complete one task. An enzyme that helps digest proteins will not be useful to break down carbohydrates.

Function Under Precise Conditions: If the temperature is too high or if the environment is too acidic or alkaline, the enzyme changes shape; this alters the shape of the active site so that substrates cannot bind to it (then the enzyme will no longer work).

Do NOT Change K_{eq}: The equilibrium constant for a reaction depends only on the difference in energy between reactants and products.

Do NOT Change ΔG for a Reaction: Enzymes only lower activation energy, but do not change the difference in energy levels between reactants and products.

Are NOT Altered or Used Up: Enzymes are not changed when they perform their function. This means that the same enzyme molecule can be used over and over again.

Hydrophilic and Hydrophobic

Substances with a special affinity for water are classified as **hydrophilic** ("water loving"). On the other hand, substances that naturally repel water, causing beads to form, are known as **hydrophobic** ("water fearing"). A hydrophilic substance is polar and often contains O–H (hydroxyl) or N–H (amino) groups that can form hydrogen bonds to water. This causes water to spread out across a hydrophilic surface in order to maximize the surface area. In contrast, hydrophobic substances usually contain C–H bonds that do not interact favorably with water causing water to bead up on a hydrophobic surface like waxed paper.

Intermolecular forces can explain many observable properties of molecules.

Water on a Hydrophilic vs. Hydrophobic Surface

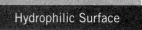

- Water "loving"
- Water adheres to the surface

- Water "fearing"
- Water repels the surface

Solutions

If a mixture is the same throughout at the particle level, we call it a **homogeneous mixture** or a **solution**. A solution forms when one substance is dissolved by another. The substance that dissolves is called the **solute**. The substance that dissolves the solute is called the **solvent**. The solute is present in a lesser amount than the solvent (clutch memorization trick: solute is also the smaller word). When the solute dissolves, it separates into individual particles, which spread evenly throughout the solvent creating a solution. Although water is the solvent for the vast majority of solutions, both solutes and solvents exist in all phases.

Process of dissolving a solute by a solvent to form a solution.

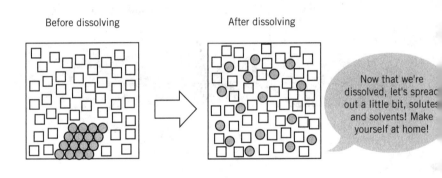

Solutions

- Have uniform distribution: Solute particles are distributed evenly thoughout the solution.
- Do NOT separate upon standing
- Cannot be separated into components via filtration, must be separated by other physical means.
- Exist in all three physical states

Molar Concentration: Molarity, M

The only measurement of concentration that you will find on the AP Exam is molarity, M, which is the number of moles (n) of solute per liter (L) of solution. This is because molarity times liters will give us the quantity of moles, which is very useful for relating the amounts of reactants and products in a balanced chemical equation. When you see a chemical symbol in brackets on the test, that means they are talking about molarity. For instance, "[K^+]" is the same as "the molar concentration (molarity) of potassium ions."

$$\text{Molarity} = M = \frac{\text{number of moles of solute }(n)}{\text{liters of solution }(L)}$$

- Molarity must be in moles per liter.
- If you are given grams of solute: Divide by molar mass of the solute to convert to moles of solute.
- If you are given mL of solution: Divide by 1000 to convert mL to L.

DO NOT worry about colligative properties in other concentration calculations (molality, %w/v , etc). These will not show up on the AP Chem Exam.

Chromatography

Chromatography is a physical method used to separate the different components in a solution. In chromatography, the sample (**analyte**) is dissolved in a particular solvent (**mobile phase**), which may be a gas or liquid. The mobile phase is then passed through another phase called **stationary phase.** That phase does not move; rather, components of the analyte travel through it at different rates depending on their polarity causing them to separate. There are different types of chromatographic techniques such as paper chromatography, column chromatography, and gas chromatography.

Chromatography exploits the different polarities of molecules in a mixture in order to separate them.

Paper Chromatography

Paper chromatography is used for identifying and separating colored mixtures like pigments. It uses paper as the stationary phase and a liquid solvent as the mobile phase. A concentrated spot of the analyte is placed on the paper and the paper is carefully dipped into a solvent (the moblile phase). The solvent rises up the paper due to capillary action and the different components of the mixture rise up at different rates and thus are separated from one another. The rate at which each component travels will be based on how attracted it is to the mobile phase versus the stationary phase. If a polar solvent is used for the mobile phase, the most polar component will travel the farthest. If a nonpolar solvent is used for the mobile phase, the most polar component will travel the least.

Next, we use the distances traveled by each component of the mixture to calculate its retention factor. The retention factor, R_f, is the ratio of the distance from the center of the spot for a given component to the distance traveled by the mobile phase, also known as the **solvent front**.

The stronger the attraction (IMFs) between the component and mobile phase is, the larger the R_f value will be.

How to Calculate R_f Values

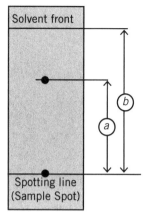

$$R_f = \frac{\text{distance traveled by the component}}{\text{distance traveled by the solvent}}$$

$$= \frac{a}{b}$$

Column Chromatography

Column Chromatography is much like paper chromatography except that the stationary phase is a solid packed into a glass column placed vertically with a downward outlet. The **analyte** is injected into the column, where it adheres to the stationary phase. After that, another solvent (called the **eluent**) is injected into the column. The process of washing a compound through a column using a solvent is known as **elution**. The components of the solution are retained by the stationary phase by varying degrees of attraction based on their polarity. Analyte molecules that are more attracted to the eluent will travel through and leave the column ("elute") first.

Column Chromatography

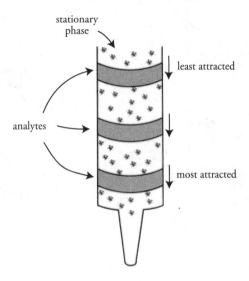

Distillation: Separation by Boiling Point

Yet another technique for separating liquid mixtures is **distillation**, which takes advantage of the different boiling points of substances in a solution in order to separate them. Distillation is commonly used to separate ethanol (the alcohol in alcoholic drinks) from water. The mixture is heated in a flask to a temperature above ethanol's boiling point, but below water's boiling point. Because ethanol has a lower point than water, it will vaporize first. The ethanol vapor is then cooled in a tube called a condenser and the pure liquid is collected as a distillate. The disadvantage of distillation is that it cannot be used to separate a mixture that contains substances with unknown or similar boiling points.

Distillation Apparatus

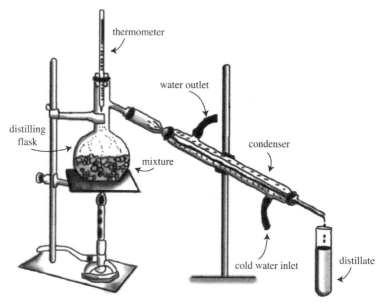

Were we able to "distill" all of that AP Chemistry material for you? Sorry, couldn't resist a good distillation joke. The kids love 'em! Onto Chapter 3!

CHAPTER 3
Stoichiometry, Precipitation Reactions, and Gas Law

In this chapter you will learn about quantitative chemistry methods that will allow you to determine the percent composition, empirical formula, and the amount of materials that are consumed and produced in a chemical reaction (stoichiometry).

Changing Matter

Chemical equations are used by chemists to track the identities and phases of the substances involved in changing matter. An equation is basically a chemical "sentence" that shows the rearrangement of atoms that occurs during a chemical change. It is like a simplified recipe for making a product. The left side of the equation contains the chemical formulas for the reactants or the "ingredients" that are being combined. The right side of the equation contains the chemical formulas for the products or the substances that are a result of the reaction. The equation may also show the phases for each substance involved in the reaction, which can help you to predict what you might observe during a reaction in the lab.

Chemical vs. Physical Changes

In general, changes in matter can be classified as either **chemical** or **physical**.

- **Chemical Changes** (also known as Chemical Reactions): Changes that involve the breaking or forming of chemical bonds
- **Physical Changes:** Changes that involve only changes in the weak intermolecular interactions, such as phase changes.

Phase changes are conversions from one state of matter to another, where the arrangement of atoms remains unchanged but the strength of the physical attractions (spacing) between molecules has changed. When a substance changes phases, its identity (and therefore its chemical formula) does not change. Changing pressure or and grinding a substance into a powder are other examples of physical changes.

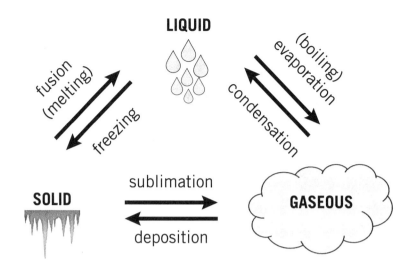

Phase changes are physical changes

But Wait! Dissolution Can be Classified as Either Physical or Chemical

When a solute is added to a solvent, the particles of the solute interact with particles of the solvent. The water will start to pull off the solute and individual solute particles will become surrounded by particles of the solvent and therefore separated from each other. When the solvent is water, the term **hydration** is often used instead of **solvation**.

In the case of molecular solutes like sucrose (table sugar), the solute particles are individual molecules and no bonds are broken during this process, so the dissolution of sugar is a physical change. However, the dissolution of soluble salts by water involves breaking the ionic bonds (between the cations and anions) and results in the formation of new (ion-dipole) interactions between the ions and the solvent (water). The magnitude of these interactions is comparable to that of covalent bond strength so the dissolution of ionic compounds are regarded as *both* a physical and chemical change.

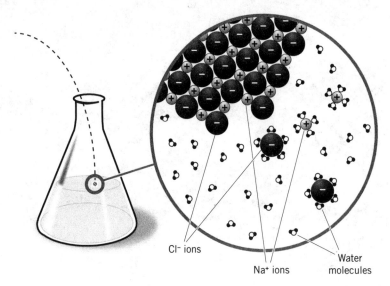

Dissolution of Ionic Salts Is Both a Chemical and Physical Change

Note that in the diagram above, the negative dipoles of the water molecules are attracted to the Na^+ cations, while the positive dipoles (hydrogen) are attracted to the Cl^- ions.

Basic Reaction Types

Below are the three basic reaction types that are not covered in other chapters of this book.

Synthesis: To Put Things Together to Make a Whole

In a synthesis reaction, two or more elements (and/or compounds) react to form a single compound. Since two things combine to form a product, some call this a combination reaction, but most chemists will use the term synthesis.

Decomposition: To break into Pieces

In a decomposition reaction, a compound is broken down into simpler components. This is the opposite of a synthesis reaction. In some cases, the reactant compound breaks down into its component elements, but more often decomposition results in smaller molecules. This process may happen in one or more steps. Because bonds are broken during decomposition, energy (usually in the form of heat, light, or electric current) is required for this process to occur.

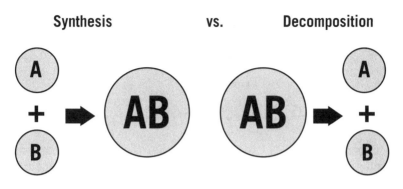

Synthesis and Decomposition are reverse chemical processes.

Combustion Reactions: Always Involve Molecular Oxygen (O_2) as a Reactant

In a combustion reaction, oxygen reacts with another element or compound (typically a hydrocarbon) to produce energy in the form of heat or light. All hydrocarbons, which contain carbon and hydrogen, react with oxygen to yield the same products, carbon dioxide (CO_2), water (H_2O), and energy. When metals combust, they form metal oxides, producing heat and light.

	Two Types of Combustion Reactions	
pe	Hydrocarbon	Metal
actants	Hydrocarbon + Molecular Oxygen	Metal + Molecular Oxygen
oducts	Carbon Dioxide + Water + Energy	Metal Oxide + Energy
ample	$CH_4 + 2\,O_2 \rightarrow CO_2 + 2\,H_2O$ + Heat	$2\,Mg + O_2 \rightarrow 2\,MgO$ + Light

 Ask Yourself...
Can the combustion of magnesium be classified as more than one type of reaction?

Did You Know?

You should understand how to recognize and predict the products of a basic synthesis, decomposition, or combustion reaction, but predicting the products of complex reactions is outside of the scope of the AP curriculum.

Basic Stoichiometry

Stoichiometry is the quantitative relationship between the amounts of reactants consumed and the amount of products formed in a chemical reaction as expressed by a balanced equation.

Balancing Chemical Equations

During a chemical reaction, atoms are **not** created or destroyed: the atoms are simply rearranged from the reactants to form the products. This is called the **Law of Conservation of Mass**. A **balanced chemical equation** is one that demonstrates that atoms are neither created nor destroyed during the reaction. To do this, you recognize that each side of the equation must contain the same number of atoms of each element involved in the reaction. If they do not, you must add coefficients (multipliers) in front of the chemical formulas to balance the atoms on each side of the reaction. Coefficients indicate how many units of each substance take part in the reaction.

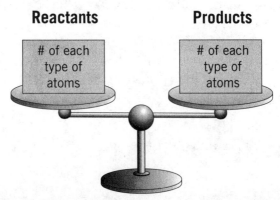

Balanced Reactions Obey the Law of Conservation of Mass

Law of Conservation of Mass

Tips for Balancing Equations

- Use coefficients and NOT subscripts to balance the number of atoms on each side of the reaction.
- If a polyatomic appears on both sides of the equation, balance it as a unit.
- When the number of atoms is even on one side and odd on the other, double the coefficients for all the atoms to make everything even.
- You can use fractional halves (1/2) as coefficients to balance diatomic molecules.
- Balance molecular hydrogen and oxygen last.
- Make sure that your final set of coefficients is reduced to the lowest whole number ratio.

Particulate Representations

One really important skill that you must demonstrate on the AP Chemistry Exam is the ability to translate the symbols of a balanced equation into meaningful diagrams of atoms, ions, and molecules. These diagrams essentially say the exact same thing as the symbolic equation, but they are another way to visualize the changes occurring at the particle level. Particulate diagrams must also obey the **Law of Conservation of Mass.**

Let's compare **Symbolic Equations** to **Particulate Representation** for the synthesis of water.

Symbolic Equation:

$2 H_2(g) + O_2(g) \rightarrow 2 H_2O(l)$ or $H_2(g) + \frac{1}{2} O_2(g) \rightarrow H_2O(l)$

Particulate Representation:

The coefficients of the balanced reaction indicate that the reaction requires twice as many H_2 molecules as O_2 molecules. Note that you CANNOT split or halve diatomic molecules when you draw a particulate diagram. Instead, you must double the other reactants and products to balance the equation.

The Mole Concept

Atoms and molecules are so tiny that they cannot be counted by direct observation. We need a bridge between these huge numbers of particles and the mass of substances, which we are able to measure directly in the laboratory. Fortunately, chemists developed a special unit called the **mole** (abbreviated mol) for the amount of substance, which provides a bridge between the number of particles and the mass of a substance.

The mole is simply a very large counting unit that is useful for counting very tiny things. Just like a dozen is useful for counting eggs, a mole is useful for counting tons of tiny particles (atoms, molecules, ions, electrons, etc). Thirteen is a baker's dozen (one extra for the baker); 6.02×10^{23} is a chemist's dozen. It makes working with lots of tiny particles much easier.

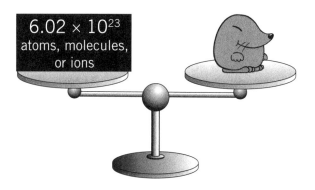

Using Subscripts to Understand Mole Ratios from Compound to Element

The subscript (the small number written to the bottom right of the element) indicates how many moles of each element are present in one mole of a particular compound. Writing the ratio for the elements in the compound can help you to determine which element is present in the larger amount of moles. For example in calcium chloride, $CaCl_2$, there are two moles of chloride per one mole of calcium. If a compound contains multiple polyatomic ions, it is important to distribute the subscript to all the atoms in the parentheses. For example in one mole of calcium nitrate, $Ca(NO_3)_2$, there are six moles of oxygen (two times three), two moles of nitrogen and one mole of calcium.

Mole Ratios in Balanced Chemical Equations

A mole ratio is the ratio comparison between the amounts of any two substances that are involved in a balanced chemical reaction. These comparisons are used as conversion factors to relate the moles of reactants to moles of product in a reaction.

Solving Stoichiometry Problems

The first step in any stoichiometry problem is to convert moles to mass and to atoms/particles. Check out this gorgeous diagram:

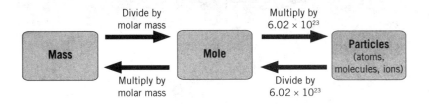

In order to determine the mass of product that will form in a given reaction, follow these three steps:

1. Determine the moles of reactant that are used.
2. Use the mole ratio from the balanced equation to convert to moles of product.
3. Convert from moles of product to mass of product by multiplying by the molar mass.

Let's do two quick examples:

$$4\ P(s) + 5\ O_2(g) \rightarrow 2\ P_2O_5$$

1. A 5.40 gram sample of phosphorus reacts with excess oxygen gas according to the above reaction.

   How many grams of P_2O_5 are produced?

$$5.40\ g\ P \times \frac{1\ mol\ P}{30.97\ g\ P} \times \frac{2\ mol\ P_2O_5}{4\ mol\ P} \times \frac{141.95\ g\ P_2O_5}{1\ mol\ P_2O_5} = 12.4\ g\ P_2O_5$$

$$2\ Al(s) + 3\ Cu(NO_3)_2(aq) \rightarrow 3\ Cu(s) + 2\ Al(NO_3)_3(aq)$$

2. Some aluminum wire is immersed in 50.0 mL of 0.100 M $Cu(NO_3)_2$, causing the above reaction to occur. How much solid copper is produced?

$$0.100\ M = \frac{n}{0.0500\ L}$$

$$n = 0.00500\ mol\ Cu(NO_3)_2 \times \frac{3\ mol\ Cu}{3\ mol\ Cu(NO_3)_2} \times \frac{63.55\ g\ Cu}{1\ mol\ Cu}$$

$$= 0.318\ g\ Cu$$

As you can see from the above examples, the path to determining the number of moles of the original reactant will be dependent on the phase it is in. We looked at solid and aqueous reactants above; we will address gases later this chapter.

Limiting Reactant Calculations

For any chemical reaction, the amount of products that can be made is dependent upon the amount of reactants that are available. Consider cooking dinner, what you can prepare is limited by the ingredients you have in your pantry and fridge (and your cooking abilities)! You can't cook what you don't have. Frequently, reactants are combined in proportions that differ from those given by the mole ratios of the balanced equation. In this case, the limiting reactant is the reactant that is used up first, which stops the reaction and limits the maximum amount of product(s) that can be made. If your recipe for two dozen cookies requires two eggs and you only have one, then you can only make one dozen cookies! All the leftover ingredients are considered to be excess reactants. The steps for determining the limiting reactant are given on the following page.

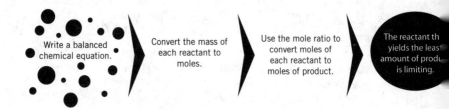

Steps for Determining the Limiting Reactant in a Reaction

Let's do an example:

$$N_2(g) + 3\, H_2(g) \rightarrow 2\, NH_3(g)$$

Ammonia is produced via the Haber process, shown above. If 2.00 g of N_2 reacts with 2.00 g of H_2, how many grams of NH_3 will be produced?

$$2.00 \text{ g } N_2 \times \frac{1 \text{ mol } N_2}{28.02 \text{ g } N_2} \times \frac{2 \text{ mol } NH_3}{1 \text{ mol } N_2} \times \frac{17.01 \text{ g } NH_3}{1 \text{ mol } NH_3} = 2.43 \text{ g } NH_3$$

$$2.00 \text{ g } H_2 \times \frac{1 \text{ mol } H_2}{2.02 \text{ g } H_2} \times \frac{2 \text{ mol } NH_3}{3 \text{ mol } H_2} \times \frac{17.01 \text{ g } NH_3}{1 \text{ mol } NH_3} = 11.3 \text{ g } NH_3$$

From the calculations above, it can be determined that the N_2 will limit, as it produces less possible product. So, 2.43 g of NH_3 will be produced.

Remember that the reactant present in the smallest amount is not necessarily the limiting reactant. The limiting reactant is the reactant that limits the amount of product made. The amount of product that can be made by the limiting reactant is called the **theoretical yield**.

Another skill that you may have to demonstrate on the exam is being able to draw (on Free Response Questions) or recognize (on Multiple-Choice Questions) the changes occurring at the particulate level before or after a reaction. This includes identifying the limiting and excess reactants in reaction.

Ask Yourself...

Can you identify the limiting and excess reactants in the particulate diagram below?

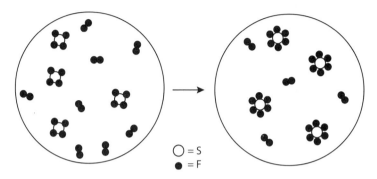

Visualizing the Limiting Reactant

Percent Yield

Percent yield is a comparison between the amount of a product an experiment produced versus how much product could have been produced (if everything went perfectly in the lab) given the amount of the limiting reactant.

Proceed Slowly

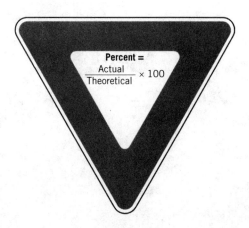

Particle Conversions Using Avogadro's Number

Avogadro's number provides a connection between the number of moles in a sample of a pure substance and the number of constituent particles (atoms, molecules, ions) of that substance.

Empirical and Molecular Formulas

There are two ways to describe the composition of a compound.

- The **empirical formula** is the simplest formula for a compound and shows the smallest whole-number ratio of the atoms present in the compound.
- The **molecular formula** indicates the ACTUAL number of atoms of each element present in one molecule of a compound. If the molecular formula is different from the empirical formula, you can determine the empirical formula by dividing each subscript by the largest whole—number factor that they have in common. For example, the molecular formula for oxalic acid is $C_2H_2O_4$. You can divide the subscripts by two to get the empirical formula for oxalic acid, CHO_2.

Percent Composition by Mass

A pure compound is always composed of the same elements in the exact same proportions by mass regardless of the amount. This is called the **Law of Definite Proportions**. The "part" is the mass of one of the elements in a sample of the compound and the "whole" is the entire mass of the compound. If you add the percent by mass of each element in a compound the total should be 100%.

To calculate the percent composition of each element in a compound apply these steps:

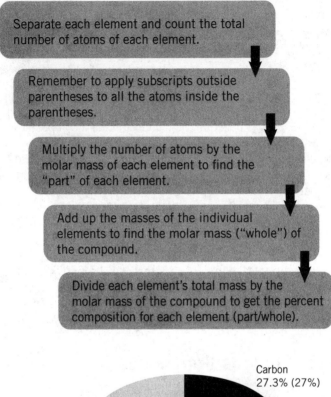

- Separate each element and count the total number of atoms of each element.
- Remember to apply subscripts outside parentheses to all the atoms inside the parentheses.
- Multiply the number of atoms by the molar mass of each element to find the "part" of each element.
- Add up the masses of the individual elements to find the molar mass ("whole") of the compound.
- Divide each element's total mass by the molar mass of the compound to get the percent composition for each element (part/whole).

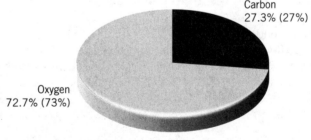

Carbon 27.3% (27%)
Oxygen 72.7% (73%)

Percent Composition of Carbon Dioxide, CO_2

Example: Determine the percent mass of each element in $Fe(NO_3)_2$.

Fe: 55.85 g × 1 =   55.85 g/mol
N: 14.01 g × 2 = + 28.02 g/mol
O: 16.00 g × 6 = + 96.00 g/mol
         Molar Mass = 179.87 g/mol

Fe: $\dfrac{55.85}{179.87}$ × 100% = 31.05% Fe

N: $\dfrac{28.02}{179.87}$ × 100% = 15.58% N

O: $\dfrac{96.00}{179.87}$ × 100% = 53.37% O

You can also work backward from mass percent to empirical formula. To do so, assume a 100. g sample, meaning any percents can also be converted directly to mass, which in turn can be converted to moles.

Once you have the percent composition of a compound, you can determine its empirical formula in just four simple steps:

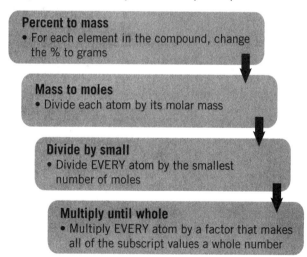

Determining the Empirical Formula from the Percentage Composition

Example: A compound is found to have 53.3% C, 11.2% H, and 35.6% O by mass. What is its empirical formula?

$$53.3 \text{ g C} \times \frac{1 \text{ mol C}}{12.01 \text{ g C}} = 4.44 \text{ mol C}$$

$$11.2 \text{ g H} \times \frac{1 \text{ mol H}}{1.01 \text{ g H}} = 11.1 \text{ mol H}$$

$$35.6 \text{ g O} \times \frac{1 \text{ mol O}}{16.00 \text{ g O}} = 2.23 \text{ mol O}$$

To determine the mole ratio, we should divide each element by the lowest number of moles—in this case, oxygen.

4.44 mol C/2.23 mol = 2 mol C
11.1 mol H/2.23 mol = 5 mol H
2.23 mol O/2.23 mol = 1 mol O

Those mole ratios become subscripts. Thus, the formula of C_2H_5O.

Finally, in a hydrocarbon combustion, you can determine the mole ratio of carbon and hydrogen in any compound by examining the amount of products produced. The trick here is to remember that there are **two** moles of hydrogen in every one mole of water!

Example: A compound containing only hydrogen and carbon is combusted, producing 2.2 g of CO_2 and 1.8 g of H_2O. What is the mole ratio of C:H in the hydrocarbon?

$$2.2 \text{ g CO}_2 \times \frac{1 \text{ mol CO}_2}{44.01 \text{ g CO}_2} \times \frac{1 \text{ mol C}}{1 \text{ mol CO}_2} = 0.050 \text{ mol C}$$

$$1.8 \text{ g H}_2\text{O} \times \frac{1 \text{ mol H}_2\text{O}}{18.01 \text{ g H}_2\text{O}} \times \frac{2 \text{ mol H}}{1 \text{ mol H}_2\text{O}} = 0.200 \text{ mol H}$$

Dividing by the smaller number, we get

0.050 mol C/0.050 mol = 1 mol C
0.200 mol H/0.050 mol = 4 mol H

Thus, the mole ratio of C:H is 1:4.

Determining the Empirical Formula from Combustion Analysis

The empirical formula can also be determined using the masses of the products of a combustion reaction (this is called **combustion analysis**). In combustion analysis, first you determine the percentage composition from the combustion data and then you use the percent composition to determine the empirical formula as outlined in the steps above.

Recall that when a hydrocarbon is combusted, the products of the reaction are always CO_2 and H_2O.

Solutions

Dissolution Diagrams (aka Particulate Diagrams)

A **dissolution diagram** (also known as a **particulate diagram**) is often used to illustrate the dissolving of an ionic solid in water and the subsequent hydration of the free ions. There are two REALLY important things that you must take into consideration when drawing the particulate diagrams for the dissolution of an ionic compound:

Size	Orientation of H_2O
The relative sizes of the cations and anions must be depicted. Recall from chapter 1 that cations become smaller and anions become larger than their neutral atoms. In most cases, the larger ion is the anion. If you are drawing a comparison for two different compounds, the ions with less energy levels must be smaller than those with more energy levels.	Water is a polar molecule, so any charged particles (such as ions) must be attracted to oppositely charged dipoles. When an ionic compound dissolves in water, the positive cations are attracted to the partially negative (oxygen) dipoles and the negative anions are attracted to the partially positive (hydrogen) dipoles on the water molecules.

Thus, when sodium chloride (NaCl) dissolves in water, it looks like this on a particulate level:

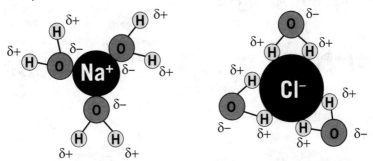

Particulate Diagram for Solvation (hydration) of Sodium Chloride

Note that at least three water molecules are included in a particulate diagram; however, in reality there are several more.

Conductivity of Ionic Solutions

When an ionic salt is dissolved in water, it will conduct electricity because the salt separates (dissociates) into ions, which are free to move. The greater the number of ions in a solution, the greater the conductivity of the solution. Therefore, a solution such as $CaCl_2$, which separates into three ions, will have a greater conductivity than a solution such as NaCl, which separates into two ions, as long as the two solutions that you are comparing have the exact same concentration.

$CaCl_2 \rightarrow Ca^{2+} + 2\ Cl^-$ (3 total ions per mole)

$NaCl \rightarrow Na^+ + Cl^-$ (2 total ions per mole)

Now, the conductivity is ultimately determined by total ionic concentration so you may need to take into consideration the total concentration of all the ions if the concentrations of the solutions are not the same. For example, when comparing solutions of 1.0 M Ca(NO$_3$)$_2$ versus 2.0 M NaCl.

1.0 M Ca(NO$_3$)$_2$ solution contains [Ca^{2+}] = 1.0 M and $\left[NO_3^-\right]$ = 2.0 M

(3.0 M free ions)

2.0 M NaCl solutions contains [Na$^+$] = 2.0 M and [Cl$^-$] = 2.0 M

(4.0 M free ions)

 Ask Yourself...
Which solution above is more conductive?

Precipitation Reactions

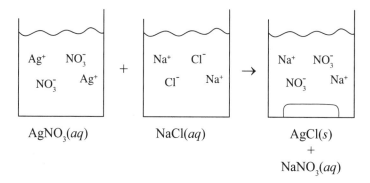

Precipitation Reactions: When Mixing Two Aqueous Solutions Produces a Solid

When two aqueous (aq) solutions are mixed and the cations and anions switch partners to form an insoluble salt(s) called a precipitate, this is called a **precipitation reaction.** Whether a precipitation reaction has occurred is determined either by observation of a precipitate in the laboratory OR by using solubility rules. Fortunately, since the AP Chemistry Exam revision in 2015, there are only three solubility rules that you must memorize and apply: salts containing alkali metals, ammonium, and nitrate are ALWAYS SOLUBLE and therefore will not form a precipitate.

Salts containing these 3 ions:

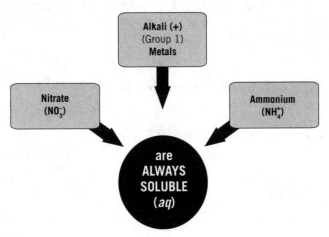

🚫 Memorization of more specific solubility rules are not required for the AP Chemistry Exam. Hooray!

Types of Equations

For reactions that occur in water, there are three general types of equations:

1. Molecular
2. Complete ionic
3. Net ionic

In a molecular equation, all reactants and products are listed in their original states. In a full ionic equation, all aqueous species are shown dissociated into their component ions. In a net ionic equation, all **spectator ions,** which are ions that do not participate in the reaction, cancel out.

Molecular: $Na_2CO_3(aq) + Ca(NO_3)_2(aq) \rightarrow 2\ NaNO_3(aq) + CaCO_3(s)$

Full Ionic: $2\ Na^+(aq) + CO_3^{2-}(aq) + Ca^{2+}(aq) + 2\ NO_3^-(aq) \rightarrow 2\ Na^+(aq) + 2\ NO_3^-(aq) + CaCO_3(s)$

Net Ionic: $CO_3^{2-}(aq) + Ca^{2+}(aq) \rightarrow CaCO_3(s)$

In the above example, Na^+ and NO_3^- are the spectator ions. On the AP Exam, you will usually be working with either the molecular or the net ionic equation.

Particulate Diagrams of Precipitation Reactions

You will also need to be able to recognize or draw which ions are limiting and which are in excess.

Gravimetric Analysis

Gravimetric analysis is a laboratory technique through which the amount of an analyte (the ion being analyzed) can be determined from measuring the mass of a precipitate followed by performing some stoichiometric calculations. The general principle is that the mass of an ion in a pure compound (the precipitate) can be determined and then used to find the mass percent of the same ion in a known quantity of an impure compound (the analyte).

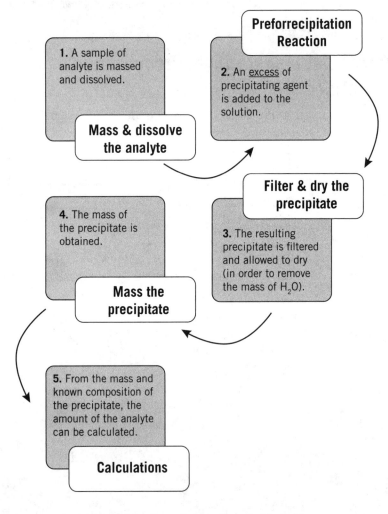

Laboratory Steps for Gravimetric Analysis

Example: A sample of a chloride salt with a mass of 0.211 g is dissolved in water and then mixed with some $AgNO_3$, causing the below reaction to occur:

$$Ag^+(aq) + Cl^-(aq) \rightarrow AgCl(s)$$

If 0.188 g of AgCl are produced, what is the mass percent of the chlorine in the original salt?

First, we have to determine the mass of chlorine in the precipitate.

$$0.188 \text{ g AgCl} \times \frac{1 \text{ mol AgCl}}{143.32 \text{ g AgCl}} \times \frac{1 \text{ mol Cl}}{1 \text{ mol AgCl}} \times \frac{35.45 \text{ g Cl}}{1 \text{ mol Cl}} = 0.0465 \text{ g Cl}$$

All of that chlorine came from the original salt, so:

$$\frac{0.0465 \text{ g Cl}}{0.211 \text{ g salt}} \times 100\% = 22.0\% \text{ Cl}$$

And that's your answer!

Kinetic Molecular Theory

The behavior of gases can be explained by a model called the **Kinetic Molecular Theory (KMT).** KMT is based on the following five postulates described here:

- Gas particles are in continuous motion, travel in a straight path, and only change direction upon collisions with other particles or the container.
- Gas particles themselves occupy no volume.
- Pressure results from collisions between the gas particles and the walls of the container.
- Collisions are elastic. The gas particles feel no attraction (IMFs) or repulsion for one another or the container.
- Average kinetic energy (KE) is directly proportional to the Kelvin temperature.

Definition of STP Conditions

- **STP** stands for **standard temperature and pressure.***
- It is defined to be 0°C (273 K) and 1 atm.
- At STP, 1 mol of any gas will occupy 22.4 L of the volume of a container. This is called **molar volume.**

Deviations From the Ideal Gas Law

In order for a gas to be ideal, its behavior must follow the Kinetic-Molecular Theory (and corresponding gas laws). Non-ideal (i.e., real) gases often deviate from this theory due to real world conditions. The ideal gas law equation functions well when the gas particles themselves do not occupy an appreciable part of the volume and when intermolecular attractions between gas particles are negligible. These criteria are satisfied under conditions of low pressure and high temperature, respectively.

On the other hand, when a gas is in a situation where its particles are crowded such as at high pressures or low temperature (i.e., high density) the behavior of gases will deviate from what is predicted by the Ideal Gas Law.

At high pressures:	At low temperatures:
Gas particles are crowded close together. The amount of empty space between the particles is reduced such that the volume of the gas particles becomes appreciable relative to the total volume occupied by the gas.	Gas particles have lower kinetic energy (KE) relative to the attractive forces between them (IMFs). The particles are less effective at overcoming IMFs upon collisions with each other.

---

*You may also recognize "STP" as the name of a very popular band from the mid-90s. They were the Stone Temple Pilots—this is standard temperature and pressure.

Stoichiometry, Precipitation Reactions, and Gas Laws

Maxwell-Boltzmann Diagrams

In a sample of gas, individual gas particles have widely different speeds; however, because of the large numbers of particles and collisions involved, the distribution and average molecular speeds are relatively constant. Maxwell-Boltzmann Diagrams are used to show the distribution of molecular speeds in a sample of gas. You need to be prepared to interpret or draw the following two types of Maxwell-Boltzmann Diagrams.

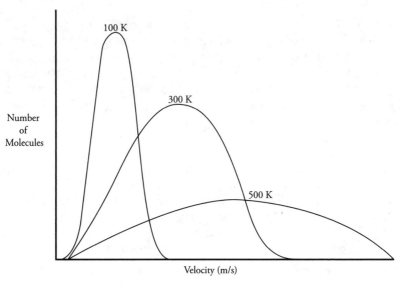

Identical Gases at Different Temperatures

If the temperature of a gas increases, its average KE increases, and therefore more molecules have higher speeds and fewer molecules have lower speeds, and the distribution shifts toward higher speeds overall (i.e., to the right). If temperature of the gas decreases, average KE decreases, and more molecules have lower speeds and fewer molecules have higher speeds, and the distribution shifts toward lower speeds overall, (i.e., to the left). The area under the curve or the total amount of all the molecules should remain the same for each temperature.

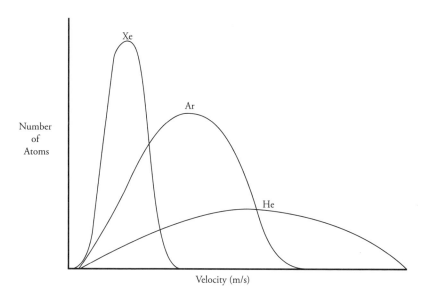

Identical Temperatures with Different Gases

At a given temperature, all gases have the same average KE. Gases composed of lighter particles have more high-speed particles with a speed distribution that peaks at relatively higher velocities (He). Gases consisting of heavier molecules have more low-speed particles and a speed distribution that peaks at relatively lower velocities (Xe).

Effusion

Effusion is the escape of gas through a small hole into an evacuated container. Think of a gas escaping through a pinhole in a balloon. At a given temperature, lighter gases (those with comparably smaller molar masses) effuse more quickly than heavier gases because the particles of the lighter gas have greater average molecular speed than the particles of the heavier gas. Yes, this is exactly what the Maxwell-Boltzmann Diagram for different gases at the same temperature demonstrates (shown above). Consequently, the rate of effusion of a gas is INVERSELY proportional to (the square root of) its molar mass. You won't have to know the equation, just the relative INVERSE relationship between mass and rate of effusion.

 Ask Yourself...
Which will effuse from a balloon faster, air or helium? How do you determine this?

Gas Laws

In order to study the behavior of gases, there are many equations that relate the different variables that describe gas behavior. In order to use these equations, it's important to look at the preferred units for each of these variables.

Unit Conversions

One variable that is used frequently to define gases is **pressure**. Pressure is defined as force divided by area. When you blow up a balloon, as you add more gas to it the pressure gradually increases and the balloon becomes "tighter."

The most common unit for pressure is atmospheres. 1 atmosphere is defined as the amount of pressure exerted by atmospheric gases at sea level. Other units you may see for pressure on the AP Exam are millimeters of mercury (mmHg) and torricellis (torr). Those terms are synonymous, and

$$1 \text{ atm} = 760 \text{ mmHg} = 760 \text{ torr}$$

💬 One final unit of pressure that you will see more often in physics classes is pascals (or kilopascals): 1 atm = 101500 Pa = 101.5 kPa.

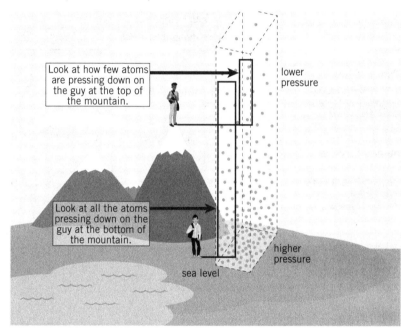

Another variable that you will use in studying gases is **temperature**. For now, the only temperature units you need to be familiar with are degrees Celsius and Kelvins. To convert between from degrees Celsius to Kelvins, simply add 273.

Finally, one more variable that comes up frequently when studying gases is **volume**. Volume is almost always measured in either milliliters or liters.

Combined Gas Law

The relationship between temperature, pressure, and volume is quantified via the Combined Gas Law:

$$\frac{P_1 V_1}{T_1} = \frac{P_2 V_2}{T_2}$$

P is pressure measured in atm, mmHg, or torr
V is volume measured in mL or L
T is temperature measured in K

When using the combined gas law, temperature MUST be in Kelvins. Volume and pressure can be in any units, so long as they are consistent (that is, if V_1 is in milliliters, V_2 must be as well.)

💬 In introductory chemistry, you may have heard of several other named gas laws.

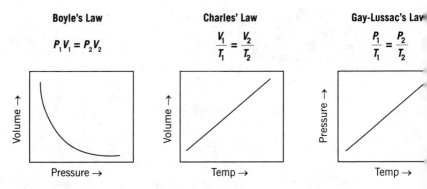

Graphical Representations of the Comparative Gas Laws

All of these gas laws are contained within the combined gas law. If any of the variables is held constant, simple cancel it out of the combined gas law before solving.

One more thing that is frequently tested on the exam is the graphical representation of pressure versus volume for gases. Simply enough, as pressure increases, volume decreases. Graphically that would look like this:

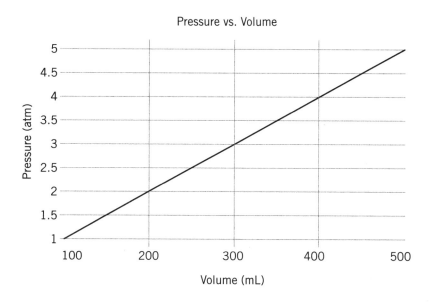

Ideal Gas Law

One of the most important equations on the AP Chemistry Exam is the Ideal Gas Law.

$$PV = nRT$$

P is pressure measured in atm
V is volume measured in L
n is moles
R is the ideal gas constant (0.0821 atmL/molK)
T is temperature measured in K

As long as you keep track of your units, the Ideal Gas Law is fairly simple to use.

Example: A 500. mL sample of nitrogen gas exerts a pressure of 795 torr at a temperature of 51°C. How many grams of gas are present in the sample?

$$P = 795 \text{ torr} \times \frac{1 \text{ atm}}{760 \text{ torr}} = 1.05 \text{ atm}$$

V = 500 mL = 0.500 L
T = 51°C + 273 = 324 K
R = 0.0821 atmL/molK

(1.05 atm)(0.500 L) = n(0.0821 atmL/molK)(324 K)
n = 0.0197 mol

Nitrogen gas is diatomic, N_2, and has a molar mass of 28.02 g/mol. 0.0197 mol × 28.02 g/mol = 0.553 g N_2 gas.

Partial Pressures

One last concept for gases that is worth being familiar with is **partial pressures**. If you have three gases in a container, adding up the pressure of all three gases will yield the total pressure in the container.

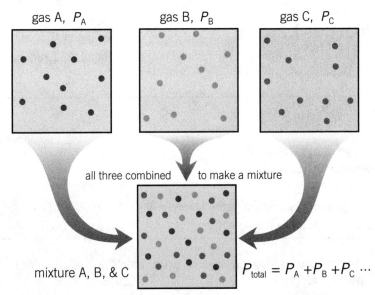

This is quantified via **Dalton's Law of Partial Pressures**:

$$P_{total} = P_A + P_B + P_C + \ldots\ldots$$

P_{total} is the total pressure in the container.
$P_A/P_B/P_C$ are the pressures of gases A, B, and C.

Another useful concept that goes along with partial pressures is mole fraction. Simply put, the mole fraction is the percentage of moles a certain gas takes up in a container.

$$X_A = \frac{n_A}{n_{total}}$$

X_A is the mole fraction of gas A.
n_A is the moles of gas A.
n_{total} is the total number of moles in the container.

The relationship between total pressure, partial pressure, and mole fraction is as follows:

$$P_A = P_T X_A$$

P_A is the pressure of gas A.
P_T is the total pressure.
X_A is the mole fraction of gas A.

Example: A sealed container is filled with 0.240 g of water vapor and 0.240 g of carbon dioxide gas. The total pressure in the container is 3.78 atm. What is the partial pressure of each gas?

First, we need to determine the mole fraction of each gas.

$$2.40 \text{ g } H_2O \times \frac{1 \text{ mol}}{18.03 \text{ g } H_2O} = 0.133 \text{ mol } H_2O$$

$$2.40 \text{ g } CO_2 \times \frac{1 \text{ mol}}{44.02 \text{ g } CO_2} = 0.0545 \text{ mol } CO_2$$

$$X_{H_2O} = \frac{0.133 \text{ mol}}{(0.133 \text{ mol} + 0.0545 \text{ mol})} = 0.707$$

Given that the total mole fractions must equal 1 (as that signifies 100%):

$$1.00 = 0.707 + P_{CO_2}$$

$$X_{CO_2} = 0.293$$

Finally:

$P_{H_2O} = X_{H_2O}(P_T)$ 	$P_{CO_2} = X_{CO_2}(P_T)$

$P_{H_2O} = (0.707)(3.78 \text{ atm})$ 	$P_{CO_2} = (0.293)(3.78 \text{ atm})$

$P_{H_2O} = 2.67 \text{ atm}$ 	$P_{CO_2} = 1.11 \text{ atm}$

As a check, 2.67 atm + 1.11 atm = 3.78 atm. That's the correct total pressure, so the math works!

When Gas is Collected over Water, Use Dalton's Law:

$$P_{Gas} = P_{total} - P_{H_2O}$$

When a gas is produced in the laboratory, the gas is often collected "over water." This means that the beaker or test tube (collection container) is filled with water and the gas is bubbled through the water and displaces some of the water. As a result, the gas and the water vapor form a mixture of gases in the test tube. This is where Dalton's Law of Partial Pressures is extremely useful. To determine the pressure of the gas, the partial pressure of the water vapor must be subtracted from the total pressure. The partial pressure of the water vapor is dependent on the temperature of the water. When you encounter this type of the problem you will either be given the partial pressure of water or a table to find the pressure using the temperature of the water.

If you see an experiment like this, use Dalton's Law.

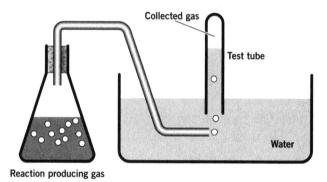

Gas Stoichiometry

Gas stoichiometry involves chemical reactions that produce gases. We can use the gas laws to help us determine the effect of temperature, pressure, and volume on the number of moles of a gas. Similar to any stoichiometry problem, the central principle is to convert moles of A to moles of B. If A and/or B are gases, you plug-in the values given for variables P, V, and T into the Ideal Gas Law to get the amount of moles (n).

Stoichiometric Calculations Gases at STP: A Shortcut!

You need to be on the lookout for reactions that take place at STP. If the reaction takes place at 273 K and 1 atm, you can use molar volume (1 mol gas = 22.4 L) as a conversion factor between volume and moles instead of using the Ideal Gas Law equation.

Write the balanced equation.
$$2\ KClO_3 \rightarrow 2\ KCl + 3\ O_2$$

Convert grams of A to moles of A.
$$5.0\ g\ KClO_3 \times \frac{1\ mol\ KClO_3}{122.55\ g\ KClO_3}$$

Multiply by the mol ratio to convert mols of A to moles of B.
$$\times \frac{3\ mol\ O_2}{2\ mol\ KClO_3}$$

Use molar volume to convert moles to liters.
$$\times \frac{22.4\ L\ O_2}{1\ mol\ O_2}$$

ICE Charts

An ICE chart is a stoichiometric graphical organizer. **ICE** stands for **Initial, Change,** and **Ending**. Using ICE charts can help you make determinations about what things will look like after a reaction has gone to completion.

Example: Some CS_2 gas at a pressure of 0.500 atm and some Cl_2 gas at a pressure of 1.00 atm are mixed into a sealed and evacuated container, causing the following reaction to occur:

$$CS_2(g) + 3\ Cl_2(g) \rightarrow CCl_4(g) + S_2Cl_2(g)$$

What will the total pressure be in the container when the reaction has gone to completion?

Because moles and pressure are directly proportional, we can use partial pressures in an ICE chart. In the "Initial" row, we put the partial pressure of each gas in. In the "Change" row, we know the reactants (CS_2 and Cl_2) will decrease, and we know the Cl_2 will decrease three times as quickly as the CS_2 due to the "3" coefficient on the Cl_2 in the balanced equation. We also know the amount of CCl_4 and S_2Cl_2 that will be created is equal to the amount of CS_2 that reacted, as all three species have a coefficient of 1.

	CS_2	Cl_2	CCl_4	S_2Cl_2
Initial	0.500	1.00	0	0
Change	$-x$	$-3x$	$+x$	$+x$
Ending				

To figure out the "Ending" row, you have to figure out which reactant is limiting. There is twice as much Cl_2 as CS_2, however, the Cl_2 is reacting three times faster. Therefore, the Cl_2 limits, meaning there will be none left at reaction completion. Solving for x:

$$1.00 - 3x = 0 \quad x = 0.333 \text{ atm}$$

Stoichiometry, Precipitation Reactions, and Gas Laws

Completing the ICE chart:

	CS_2	Cl_2	CCl_4	S_2Cl_2
Initial	0.500	1.00	0	0
Change	$-x$	$-3x$	$+x$	$+x$
Ending	0.177	0	0.333	0.333

The total pressure in the container will be a sum of the partial pressures, so 0.177 atm + 0.333 atm + 0.333 atm = 0.843 atm.

Density and Molar Mass

We can combine the concepts of density (measured in g/L) and the Ideal Gas Law to come up with an equation that relates density and molar mass. The steps to get there aren't important, but you should know the following equations.

$$MM = DRT/P \text{ and } D = P(MM)/RT$$

Where D is density (in g/L)
MM is molar mass (in g/mol)
R is the ideal gas constant
T is temperature (in K)
P is pressure (in atm)

CHAPTER 4
Thermochemistry

All chemical reactions involve either the breaking of old bonds or the formation of new ones. Most reactions involve both. During this process, there will be a transfer of energy between the reaction and its surroundings. Studying this energy exchange is called thermochemistry.

Thermal Energy 🛑

Temperature and Heat

Temperature measures the average amount of energy in a substance, and it has several scales.

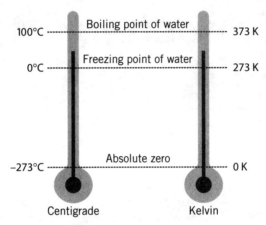

- 0°C + 273 = Kelvins.
- Absolute zero = 0 Kelvin = all molecular motion stops.
- Temperature and heat are not the same thing!

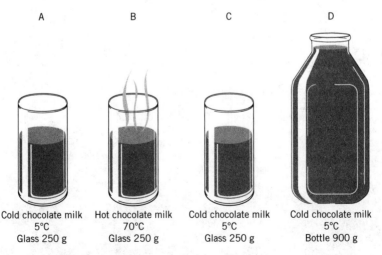

A has less thermal energy then B.   C has less thermal energy then D.

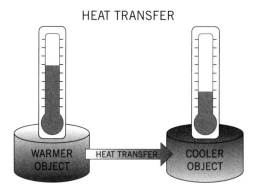

Heat is the flow of energy between two objects, and it is measured in Joules.

When two objects with different temperatures touch, heat flows from the one at the higher temperature to the one at the lower temperature. This can be as a result of a physical process (such as when a hot metal is dropped in water) or as a chemical reaction. Heat will always flow between objects until they reach the same temperature, a state that is called thermal equilibrium.

Specific Heat

Specific heat is defined as the amount of energy necessary to raise one gram of a substance by one degree Celsius (or one Kelvin).

In the above diagram, the student in the black feels a lot hotter than the student in white. This is because black has a LOWER specific heat, so it's easier to change the temperature of the black shirt.

Water has a specific heat of 4.18 J/g°C. That means for every 4.18 Joules of energy added to it, one gram of water would increase temperature by one degrees Celsius.

The formula relating energy, specific heat, and temperature is:

$$q = mc\Delta T$$

q = heat (measured in Joules)
m = mass (measuredd in grams)
c = specific heat (measured in J/g°C)
ΔT = temperature change (measured in degrees Celsius)

Let's look at a quick example:

Ethanol has a specific heat of 2.46 J/g°C and a density of 1.0 g/mL. A sample of ethanol is initially at a temperature of 22.5°C. If 5500 J of heat are added, what would the final temperature of the ethanol be?

$$5500 \text{ J} = (50.0 \text{ g})(2.46 \text{ J/g°C})\Delta T$$
$$\Delta T = 44.7°C$$
$$22.5°C + 44.7°C = 67.2°C$$

There are other ways to relate heat to temperature as well. There's molar heat capacity:

$$q = nc\Delta T$$

q = heat (measured in Joules)
n = moles (measured in moles)
c = specific heat (measured in J/g°C)
ΔT = temperature change (measured in degrees Celsius)

And there's also just plain old heat capacity, which is often used with calorimeters that have a constant volume:

$$q = C\Delta T$$

q = heat (measured in Joules)
C = heat capacity (measured in kJ/°C)
ΔT = temperature change (measured in degrees Celsius)

If you're ever confused as to which to use, make sure you look at the units on the heat capacity.

Work It [Let Me Work It] 💬

Another way energy can transfer between two objects, other than as heat, is through work.

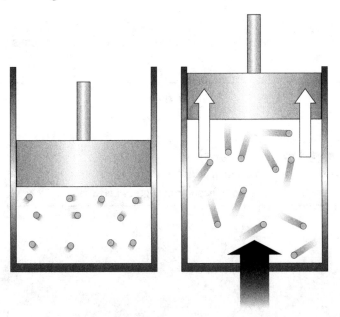

Within the confines of the AP Chemistry Exam, work occurs when a gas expands and exerts force on a movable object, such as a piston.

Enthalpy 🔔

Breaking bonds requires energy. Forming bonds releases energy.

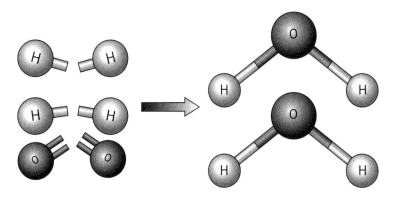

An exothermic reaction releases energy into its surroundings, increasing temperatures.

An endothermic reaction absorbs energy from its surroundings, decreasing temperatures.

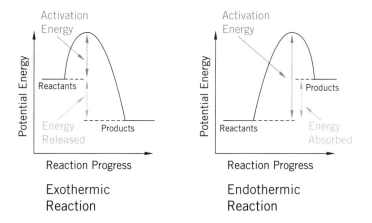

Exothermic Reaction

Endothermic Reaction

The total energy change that occurs over the course of a reaction is called the enthalpy of reaction, usually abbreviated $\Delta H°_{rxn}$.

1 "mole of reaction" can be converted to moles of any species using coefficients.

$$N_2 + 3\,H_2 \rightarrow 2\,NH_3(g) \quad \Delta H_0 = -92 \text{ kJ/mol}_{rxn}$$

How much heat will be released when 1.00 g of hydrogen reacts with excess nitrogen to create ammonia via the above reaction?

$$1.00 \text{ g } H_2 \times \frac{1 \text{ mol } H_2}{2.00 \text{ g } H_2} \times \frac{1 \text{ mol}_{rxn}}{3 \text{ mol } H_2} \times \frac{-92 \text{ kJ}}{1 \text{ mol}_{rxn}} = -15 \text{ kJ}$$

Specific enthalpy types that are good to know would be:

1. **Enthalpy of Formation.** Amount of energy released or absorbed when one mole of a substance is created from elements in their standard stares. For methanol, CH_3OH, the reaction would be $C(s) + 2\,H_2(g) + \frac{1}{2}\,O_2(g) \rightarrow CH_3OH(g)$. For diatomics like the H_2 and O_2 above, the elements must be represented in their standard state.

2. **Enthalpy of Combustion.** The amount of energy that is released when one mole of a substance is combusted in oxygen under standard conditions. For methanol, this would be $CH_3OH(l) + \frac{3}{2}\,O_2(g) \rightarrow CO_2(g) + 2\,H_2O(l)$, and $\Delta H°_{comb} = -704 \text{ kJ/mol}_{rxn}$. Enthalpy of combustion is always negative, as combustion reactions always emit energy.

3. **Enthalpy of Solution.** There are three steps in the dissolution process when an ionic solute dissolves in water. We'll use the dissolution of sodium chloride in our example.
    i) Breaking bonds
    The part of the process requires energy, and is thus always endothermic.

```
Na⁺— Cl⁻ — Na⁺— Cl⁻              Na⁺         Cl⁻
 |     |      |     |      →
Cl⁻ —Na⁺— Cl⁻ — Na⁺                          Cl⁻
                                  Na⁺
```

ii) Separating solvent ions
The water molecules spread apart. This requires the weakening of IMFs, and it is also endothermic.

iii) New attractions form
The dissociated ions will form attractions to the dipoles in water. The formation of these attractions released energy, and thus this step is always exothermic.

The sum of the enthalpies of all three steps would give you the enthalpy of solution for the salt.

Thermochemistry

Calculating $\Delta H°_{rxn}$ 🛑

There are four different methods that can be used to calculate $\Delta H°_{rxn}$.

Calorimetry

We cannot track the temperature change of a reaction directly, but by tracking the temperature change of its surroundings, we can work backwards to determine the enthalpy change for the reaction.

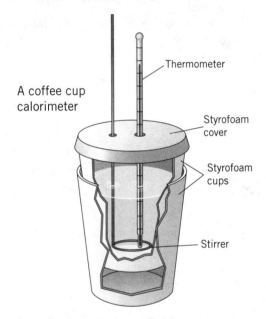

A coffee cup calorimeter

- Thermometer
- Styrofoam cover
- Styrofoam cups
- Stirrer

A coffee cup calorimeter, like the one above, is often used to track energy flow by measuring the temperature change of the solution inside.

Heat being gained leads to a positive q value, while heat being emitted leads to a negative q value.

All calorimetry problems can be done with three steps:

1. Calculate the heat gained or lost by the water.
2. Flip the sign to determine the heat gained or lost by the reaction.
3. Divide the heat gained or lost by the reaction by the moles of reaction.

Let's try an example!

$$H^+(aq) + OH^-(aq) \rightarrow H_2O(l)$$

When 100. mL of 1.0 M NaOH is mixed with 150. mL of 1.0 M HCl via the above reaction, the temperature change of the resulting solution increases by 5.51°C. Assuming the final solution has a density and specific heat identical to pure water, what is the enthalpy of reaction?

1. $q = mc\Delta T$
   $q = (250. \text{ g})(4.18 \text{ J/g°C})(5.51°C)$
   $q = 5760 \text{ J}$

2. $q_{rxn} = -q_w$
   $q_{rxn} = -5760 \text{ J}$

3. $\Delta H_{rxn} = \dfrac{q_{rxn}}{n_{rxn}}$

$$1.0 \, M = 1.00 \, M = \dfrac{n}{0.100 \text{ L}}$$

$$n = 0.100 \text{ mol NaOH} \times \dfrac{1 \text{ mol}_{rxn}}{1 \text{ mol NaOH}} = 0.100 \text{ mol}_{rxn}$$

$$\Delta H_{rxn} = \dfrac{-5760 \text{ J}}{0.100 \text{ mol}_{rxn}} = -57600 \text{ J/mol}_{rxn} = -57.6 \text{ kJ/mol}_{rxn}$$

It's important that you watch your significant figures when doing calorimetry problems, as the amount of energy transfer is usually a very large number that needs to be rounded carefully to maintain the correct number of sig figs.

Also, during any calorimetry experiment, heat will not only transfer between the reaction and the solution. Some of that heat will be lost to the air and to the container the reaction is occurring in, meaning experimentally calculated ΔH_{rxn} will usually be artificially low.

Thermochemistry

Enthalpies of Formation

If you are given a data table that shows the enthalpy of formation, sum up all the enthalpies of formation for the products, and then subtract from those all of the enthalpies of formation for the reactants.

Using the data provided, find the enthalpy of reaction for the following reaction:

$$2\ NO_2(g) + 7\ H_2(g) \rightarrow 2\ NH_3(g) + 4\ H_2O(l)$$

Substance	$\Delta H_f°$ (kJ/mol)
NO_2 (g)	34.0
NH_3 (g)	−45.9
H_2O (l)	−285.8

2(−45.9) + 4(−285.8) − (2(34.0) + 7(0)) = −1,167 kJ/mol

Bond Enthalpies

If you are given a table of bond enthalpies, you can use these to calculate the total $\Delta H°_{rxn}$. Simply assign broken bonds a positive value, formed bonds a negative value, and sum all of them up.

When doing this, make sure you account for both any coefficients present in the reaction, but also the number of bonds present within each species.

Using the provided bond energy values, calculate the enthalpy change, ΔH_{rxn}, that occurs in the following reaction.

$$CH_4(g) + 2\ O_2(g) \rightarrow CO_2(g) + 2\ H_2O(g)$$

Bond	Enthalpy (kJ/mol)
C–H	414
O=O	498
C=O	799
O–H	465

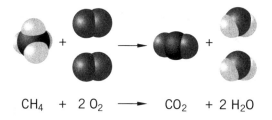

CH₄ + 2 O₂ ⟶ CO₂ + 2 H₂O

4(414) + 2(498) + 2(–799) + 4(–465) = –806 kJ/mol$_{rxn}$

Hess's Law

If you can manipulate several reactions with known enthalpy values in such a way that when they are combined, you get another reaction, you can also calculate the enthalpy of that reaction. There are three rules:

1. If you flip a reaction, the sign on the enthalpy value also flips.
2. If you multiply or divide all of the coefficients in a reaction by an integer, the enthalpy is also multiplied or divided by the same integer.
3. If you sum multiple reactions, the enthalpy values of those reactions will also be summed.

Let's see this in action. Find the enthalpy of reaction for the combustion of propane using the data provided below.

$$C_3H_8(g) + 5\ O_2(g) \rightarrow 3\ CO_2(g) + 4\ H_2O(l)$$

C (graphite) + $O_2(g)$ → $CO_2(g)$ $\quad\quad\quad\quad\quad\quad\quad\quad$ ΔH = –394 kJ/mol

$H_2O(l)$ → $H_2(g)$ + ½ $O_2(g)$ $\quad\quad\quad\quad\quad\quad\quad\quad$ ΔH = 286 kJ/mol

$C_3H_8(g)$ → 3 C (graphite) + 4 $H_2(g)$ $\quad\quad\quad\quad\quad$ ΔH = 105 kJ/mol

First, flip the middle reaction and multiply it by four:

4 $H_2(g)$ + 2 $O_2(g)$ → $H_2O(l)$ $\quad\quad\quad$ ΔH = (–286)4 = –1,144 kJ/mol

Then, multiply the top reaction by three:

3 C (graphite) + 3 $O_2(g)$ → 3 $CO_2(g)$ $\quad$ ΔH = –394(3) = –1,182 kJ/mol

Thermochemistry

Finally, add up all three reactions:

$$\begin{aligned}
&\cancel{3\text{ C (graphite)}} + 3\text{ O}_2(g) \rightarrow 3\text{ CO}_2(g) &&\Delta H = -1{,}182 \text{ kJ/mol} \\
&+ \cancel{4\text{ H}_2(g)} + 2\text{ O}_2(g) \rightarrow \text{H}_2\text{O}(l) &&+\Delta H = -1{,}144 \text{ kJ/mol} \\
&+ \text{C}_3\text{H}_8(g) \rightarrow \cancel{3\text{ C (graphite)}} + \cancel{4\text{ H}_2(g)} &&+\Delta H = 105 \text{ kJ/mol}
\end{aligned}$$

$\text{C}_3\text{H}_8(g) + 5\text{ O}_2(g) \rightarrow 3\text{ CO}_2(g) + 4\text{ H}_2\text{O}(l)$ $\quad \Delta H = -2{,}221 \text{ kJ/mol}$

> 🚫 **Exclusion Alert: State Functions**
>
> Your teacher and/or textbook may discuss "state functions" while studying enthalpy. This terminology and the concepts underlying it are no longer required on the AP Exam.

Heating & Cooling Curves

If we were to add heat to ice until it melted into liquid, and then kept adding heat until the liquid boiled into a gas, you would get a "curve" that looks like the below:

Heating Curve for Water at 1.00 atm Pressure

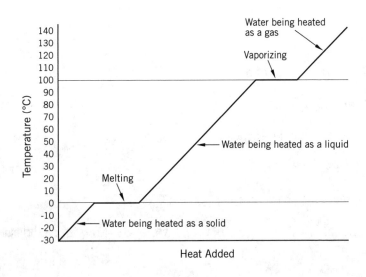

The two flat portions of the graph occur when the ice is melting, or the water is boiling.

All energy during phase changes goes into overcoming IMFs, so when substances are changing phase, their temperature remains constant!

Heat of fusion (sometimes called enthalpy of fusion) describes how much energy is required to melt a substance.

Heat of vaporization (sometimes called enthalpy of vaporization) describes how much energy is required to boil a substance.

A 5.00 g sample of ice at 0°C (ΔH_{fusion} = 6.02 kJ/mol) is heated until it melts fully. The liquid water (c = 4.18 J/g°C) is then heated until it reaches a temperature of 20.0°C. How much heat is added during the entire process?

First, deal with heating the ice during the phase change:

$$5.00 \text{ g } H_2O \times \frac{1 \text{ mol } H_2O}{18.0 \text{ g } H_2O} = 0.278 \text{ mol } H_2O$$

$$\Delta H_{fusion} = \frac{q}{n} \qquad 6.02 \text{ kJ/mol} = \frac{q}{0.278 \text{ mol}} \qquad q = 1.67 \text{ kJ}$$

Then, deal with the heat added to raise the temperature of the water:

$$q = (5.00 \text{ g})(4.18 \text{ J/g°C})(20.0°C)$$

$$q = 418 \text{ J}$$

The units don't match, so we need to make sure to address that when adding the two heats together:

$$q = 1.67 \text{ kJ} + 0.418 \text{ kJ} = 2.09 \text{ kJ}$$

Thermochemistry

Entropy !

Entropy is the tendency to become less organized over time. Yes, it's a phenomenon that could be about your school locker as the academic year goes on, but it's about chemistry.

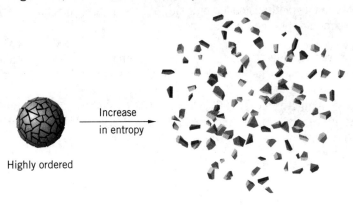

Highly ordered → Increase in entropy → More disordered

Entropy change is symbolized by ΔS, where a positive ΔS indicates an increase in disorder, and a negative ΔS indicates a decrease in disorder. Being able to predict the entropy change of a chemical reaction is a vital skill.

1. **Phase changes.**
   $NaHCO_3(s) + CH_3COOH(aq) \rightarrow$
   $CO_2(g) + H_2O(l) + Na^+(aq) + CH_3COO^-(aq)$

   A gas is produced and a solid is consumed, leading to a positive ΔS.

2. **Changing number of gas molecules.**
   4 gas molecules → 3 yields a negative ΔS.
   $2\ H_2(g) + 2\ NO(g) \rightarrow N_2(g) + 2\ H_2O(g)$

3. **Changing concentrations.**
   Lower concentrations have more "spread out" solutes and are thus more disordered.

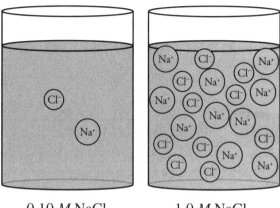

0.10 M NaCl   1.0 M NaCl

 4. **Temperature change.**
   As temperature increases, there is a wider range of potential velocities for the particles in the gas, meaning the entropy will increase.

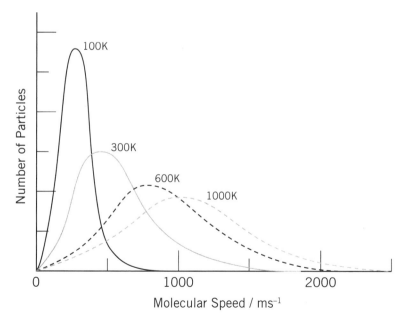

Thermochemistry

Absolute Entropy

A substance that has 0 absolute entropy would be completely ordered. It is thus impossible for the absolute entropy of a substance to be negative. If you are given a table of absolute entropy values, you can calculate the entropy change for a reaction by summing up the absolute entropy values for all the products, and then subtracting the sum of all the absolute entropy values of all the reactants. This works identical to the "enthalpy of formation" example found earlier in this chapter.

Reaction Favorability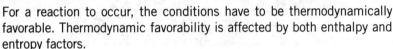

For a reaction to occur, the conditions have to be thermodynamically favorable. Thermodynamic favorability is affected by both enthalpy and entropy factors.

Exclusion Statement

On the previous exam, "thermodynamically favorable" was called "spontaneous" and "thermodynamically unfavorable" was called "non-spontaneous." On the AP Chemistry Exam, the terms thermodynamically favorable and thermodynamically unfavorable will be used exclusively.

Gibbs Free Energy

For a reaction to occur, one (or both) of two things must be true.

1. A reaction releases energy (negative ΔH).
2. A reaction leads to more disorder (positive ΔS).

These two variables are combined into a concept called Gibbs free energy via the following equation.

$$\Delta G = \Delta H - T\Delta S$$

ΔG = Gibbs free energy (in kJ/mol)
ΔH = enthalpy change (in kJ/mol)
T = temperature (in Kelvins)
ΔS = entropy change (in J/mol * K)

- Reactions are only favorable when the ΔG value is negative.
- A reaction with a negative ΔG value is called exergonic.
- A reaction with a positive ΔG value is called endergonic.
- When doing Gibbs calculations, make sure your units line up correctly!

2 $CH_3OH(g)$ + 3 $O_2(g)$ → 2 $CO_2(g)$ + 4 $H_2O(l)$
$\Delta H° = -1538$ kJ/mol, $\Delta S° = -387$ J/mol·K

The combustion of ethanol has the enthalpy and entropy change values described above. Calculate $\Delta G°$ for this reaction under standard conditions (298 K). Is the reaction favored at that temperature?

$$\Delta G° = \Delta H° - T\Delta S°$$

First, we have to make the units match:

$$387 \text{ J/mol·K} = 0.387 \text{ kJ/mol·K}$$

Then we solve!

$\Delta G° = -1,538$ kJ/mol $-$ (298 K)(0.387 kJ/mol·K)
$\Delta G° = -1,653$ kJ/mol

The reaction is favored because $\Delta G°$ is negative.

Thermochemistry

Phase changes are physical processes, and as such, $\Delta G° = 0$ for any phase change. That sort of information could be useful in problems like the one below.

Acetone has a heat of vaporization (ΔH_{vap}) of 29.1 kJ/mol and an entropy of vaporization (ΔS_{vap}) of 88.4 J/mol·K. What is the boiling point of acetone?

Again, first we have to make the units match:

$$88.4 \text{ J/mol·K} = 0.0884 \text{ kJ/mol·K}$$

Then, we use our Gibbs equation and set $\Delta G° = 0$.

$$\Delta G° = \Delta H° - T\Delta S°$$

$0 = 29.1 \text{ kJ/mol} - T(0.0884 \text{ kJ/mol·K})$
$T(0.0884 \text{ kJ/mol·K}) = 29.1 \text{ kJ//mol}$
$T = 329 \text{ K}$

Temperature Effects

Temperature directly effects the entropy change in a reaction. The table below shows the relationship between all three variables that need to be considered when determining reaction favorability.

ΔH	ΔS	Favored?
+	+	At high temperatures
+	−	Never
−	+	Always
−	−	At low temperatures

For phase changes, the value for ΔG is always zero because no chemical reaction is occurring.

Overcoming Unfavorable ΔG Values

If a reaction is not favored, it can still be forced to occur under certain conditions.

1. **Adding electricity.** If an electrical current is run through a system, that can cause a normally unfavorably reaction to occur.
2. **Exposing a reaction to radiation.** Electromagnetic radiation has an inherent amount of energy based on its frequency. Some very common chemical processes, such as photosynthesis, require the input of additional light/radiation in order to occur.

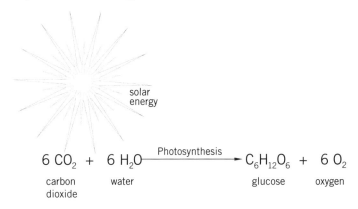

3. **Coupling an unfavorable reaction with a favorable one.** The conversion of ATP to ADP is a good example of this.

* **Just because a reaction is favored does not mean it is fast!** *
Favored ≠ Fast!

A reaction that is favored but does not seem to be occurring at a measurable rate is said to be under kinetic control.

Reactions that are favored and occur as expected are called under thermodynamic control.

CHAPTER 5

Equilibrium and the Solubility Product Constant

We've been considering chemical reactions as one-way streets so far: reactants come together, react, and make products. It's time to turn this one-way street into a two-way street and consider the possibility that products can also turn into reactants. Equilibrium is the study of the balance between forward and reverse directions on this two-way street.

Equilibrium Concepts 🛑

Most chemical processes are **reversible**. In other words, not only do reactants react to form products, but products can react to form reactants. A reaction is said to be at equilibrium when the rate of the forward reaction is equal to the rate of the reverse reaction. The concentrations of reactants and products will be constant once equilibrium is achieved, but that doesn't mean that the reactions stop! Since change continues to occur on a microscopic scale, this is called a **dynamic equilibrium**.

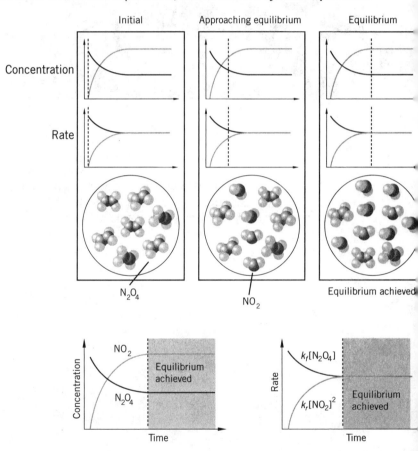

The Equilibrium Constant

The relationship between the concentrations of reactants and products in a reaction at equilibrium is given by the equilibrium expression.

The Equilibrium Expression

For the reaction
$$aA + bB \rightleftharpoons cC + dD$$

$$K_{eq} = \frac{[C]^c [D]^d}{[A]^a [B]^b}$$

Let's take a look at how K_{eq} is written from a balanced reaction:

$$2\ CO(g) \rightleftharpoons CO_2(g) + C(s)$$

1. Products on the top

$$K_{eq} = \frac{[CO_2]}{[CO]^2}$$

2. Coefficients become exponents

A few things to keep in mind here:

1. Products are in the numerator and reactants are in the denominator [CO_2 on the top, CO on the bottom]
2. Coefficients become exponents [**2** CO in reaction becomes $[CO]^2$ in K_{eq}]
3. Solids [C(s) in this case] and liquids are not included
4. No units for K_{eq}

Equilibrium and the Solubility Product Constant

K_c vs. K_p

If only aqueous species are involved in the equilibrium expression, K_{eq} may also be called K_c. If only gases are involved in the equilibrium expression, K_{eq} may also be called K_p. In the case of K_p, the values entered into the expression will be partial pressures, and the corresponding units on them are atmospheres.

> ⊘ For the AP Chemistry Exam, you do not need to know how to convert between K_c and K_p.

K_{eq} and Reaction Favorability

The size of K_{eq} is also related to the favorability of the reaction and the relative amounts of products and reactants at equilibrium. It tells us the same thing as the sign of $\Delta G°$.

The Relationship between K_{eq} and $\Delta G°$		
$\Delta G° = -RT \ln K_{eq}$		
$K_{eq} < 1$ (small) and $\Delta G° > 0$	$K_{eq} = 1$ and $\Delta G° = 0$	$K_{eq} > 1$ (large) and $\Delta G° < 0$
(reactants heavy, products light)	(balanced)	(products heavy, reactants light)
non-favorable forward reaction; reactants favored at equilibrium	comparable amounts of products and reactants at equilibrium	favorable forward reaction; products favored at equilibrium

Manipulating K_{eq}

Much as Hess's Law allows for the determination of the enthalpy change for a reaction given the enthalpy change values for similar reactions, you can determine the equilibrium constant of a reaction by manipulating similar reactions with known equilibrium constants. (Note: flip back to Chapter 4 if you need a refresher on Hess's Law.) However, the rules for doing so are different than the rules for enthalpy values. Let's look at some examples here:

Flipping a Reaction	Multiplying by a Coefficient	Add Two Reactions Together
Take the reciprocal of the equilibrium constant to get the new equilibrium constant.	Raise the equilibrium constant to that power to get the new constant.	Multiply the equilibrium constants of those reactions to get the new constant.

Reaction 1a:

$N_2(g) + 3\ H_2(g) \rightleftharpoons 2\ NH_3(g)$

K_{eq} of reaction 1 =

$$\frac{[NH_3]^2}{[N_2][H_2]^3} = 0.0001$$

Reaction 1b:

$2\ NH_3 \rightleftharpoons N_2 + 3\ H_2$

K_{eq} of reaction 1b =

$$\frac{[N_2][H_2]^3}{[NH_3]^2} =$$

$$\frac{1}{0.0001} = 10{,}000$$

Reaction 2a:

$H_2(g) + \frac{1}{2} O_2(g) \rightleftharpoons H_2O(g)$

K_{eq} of reaction 2a =

$$\frac{[H_2O]}{[H_2][O_2]^{\frac{1}{2}}} = 300$$

Reaction 2b:

$2\ H_2(g) + O_2(g) \rightleftharpoons 2\ H_2O(g)$

Or:

$2 \times [H_2(g) + \frac{1}{2} O_2(g)$

$\rightleftharpoons H_2O(g)]$

K_{eq} of reaction 2b =

$$\frac{[H_2O]^2}{[H_2]^2[O_2]} =$$

$300^2 = 90{,}000$

Reaction 3a:

$N_2(g) + O_2(g) \rightleftharpoons 2\ NO(g)$

K_{eq} of reaction 3a = 0.25

Reaction 3b:

$2\ NO(g) + O_2(g) \rightleftharpoons 2\ NO_3(g)$

K_{eq} of reaction 3b = 4000

Reaction 3c, from adding reactions 3a + 3b:

$N_2(g) + 2\ O_2(g) \rightleftharpoons 2\ NO_3(g)$

K_{eq} of reaction 3c: $0.25 \times 4000 = 1000$

Equilibrium and the Solubility Product Constant

Equilibrium Stressors

At equilibrium, the rates of the forward and reverse reactions are equal. A "shift" in a certain direction means the rate of the forward or reverse reaction increases, so that a net reaction is observed in that direction.

Le Châtelier's Principle

Le Châtelier's Principle states that whenever a stress is placed on a system at equilibrium, the system will shift in response to that stress to re-establish equilibrium. If the forward rate increases, we say the reaction has shifted right, which will create more products. If the reverse rate increases, we say that the reaction has shifted left, which will create more reactants.

Let's consider equilibrium shifts in the Haber process for making ammonia: $N_2(g) + 3\ H_2(g) \rightleftharpoons 2\ NH_3(g)$.

Concentration

When the concentration of a reactant or product is increased, the reaction will shift in the direction that allows it to use up the added substance. If N_2 or H_2 is added, the reaction shifts right. If NH_3 is added, the reaction shifts left.

Check out this graph of concentration vs. time when H_2 is added.

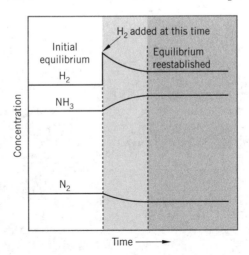

Since the text mentions rate, this additional graph may also be helpful for you as you study.

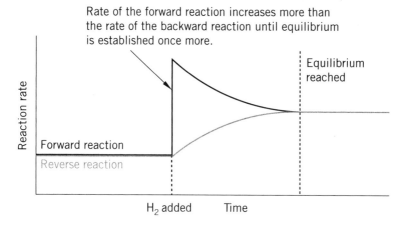

When the concentration of a species is decreased, the reaction will shift in the direction that allows it to create the substance that has been removed. If N_2 or H_2 is removed, the reaction shifts left.

Pressure ❗

When the external pressure on a system in increased, that will increase the partial pressure of all the gases inside the container and cause a shift to the side with fewer gas molecules.

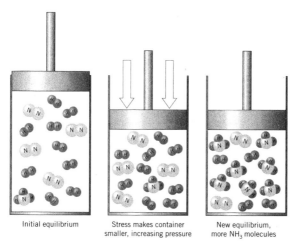

$$N_2(g) + 3\ H_2(g) \rightleftharpoons 2\ NH_3(g)$$

Equilibrium and the Solubility Product Constant

In the **Haber process**, there are four gas molecules on the left hand-side of the reaction (one N_2 molecule and three H_2 molecules) and two on the right (two NH_3 molecules), so an increase in partial pressures of the gases (by making the container smaller, for example) would cause a shift to the right. Decreasing the partial pressures of all gases (by making the container bigger) would have the opposite effect and cause a shift to the left.

In addition to changing volume, the partial pressures of gases can be changed by adding another gas to the container while maintaining the total pressure, even if that gas does not participate in the reaction.

If we were to add some helium gas to a container where the Haber process was in equilibrium and the pressure is held constant, it would decrease the mole fraction of all the gases participating in the reaction (since the helium now represents at least some percentage of the total gases). A decreased mole fraction causes a decreased partial pressure, which again causes a shift to the left. However, if when adding the helium the total pressure were to increase, no shift would occur, as the partial pressure of the helium would just be added to the existing (and unchanging) partial pressures of the gases participating in the reaction.

Note that for any equilibrium, there must be a different number of gas molecules on each side of the equilibrium for a pressure change to cause any kind of shift. If the reaction has the same number of gas molecules on both sides, the changes above (changing the size of the container or adding an inert gas) would cause no shift at all.

Temperature

There's a trick to figure out what happens when the temperature changes. Adding or removing heat is just like adding or removing a reactant or product. We just need to re-write the equation to include the heat energy on the side that it would be present on, according to the enthalpy change.

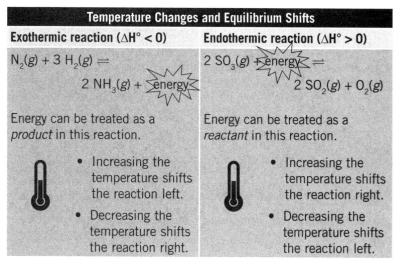

Temperature Changes and Equilibrium Shifts	
Exothermic reaction ($\Delta H° < 0$)	**Endothermic reaction ($\Delta H° > 0$)**
$N_2(g) + 3 H_2(g) \rightleftharpoons 2 NH_3(g) +$ energy	$2 SO_3(g) +$ energy $\rightleftharpoons 2 SO_2(g) + O_2(g)$
Energy can be treated as a *product* in this reaction.	Energy can be treated as a *reactant* in this reaction.
• Increasing the temperature shifts the reaction left. • Decreasing the temperature shifts the reaction right.	• Increasing the temperature shifts the reaction right. • Decreasing the temperature shifts the reaction left.

Unlike the other equilibrium stressors, changes in temperature also change the value of the equilibrium constant. A shift to the left decreases the value of the equilibrium constant, while a shift to the right results in an increase.

Dilutions

One last type of shift is caused by dilution, which can be observed in aqueous equilibria. The effect of dilution is analogous to increasing the size of the container in a gas-phase equilibrium.

Dilution and Equilibrium Shifts	
$Fe^{3+}(aq) + SCN^-(aq) \rightleftharpoons FeSCN^{2+}(aq)$	
Adding water causes a shift to the side with **more** aqueous species (in the above example, the left-hand side).	**Removing** water (by evaporation, perhaps) causes a shift to the side with **fewer** aqueous species (in the above example, the right-hand side).

Experimentally Measurable Shifts !

Some equilibrium shifts can be observed very easily.

Color

$$Fe^{3+}(aq) + SCN^-(aq) \rightleftharpoons FeSCN^{2+}(aq)$$

Fe^{3+} is a pale yellow color, and SCN^- is colorless. However, $FeSCN^{2+}$ is a dark red-brown color. As this equilibrium shifts to the right, making more $FeSCN^{2+}$, we observe the color getting darker and darker.

Temperature

$$Co^{2+}(aq) + 4\ Cl^-(aq) + heat \rightleftharpoons CoCl_4^{2-}(aq)$$

Co^{2+} is a light pink color. When it is heated in the presence of Cl^- ions, the dark blue $CoCl_4^{2-}$ is formed.

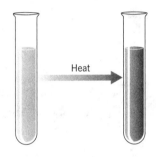

pH

$$Zn(OH)_2(s) \rightleftharpoons Zn^{2+}(aq) + 2\ OH^-(aq)$$

If H^+ (in the form of HCl, for example) is added to a saturated solution of $Zn(OH)_2$ with solid $Zn(OH)_2$ on the bottom, more $Zn(OH)_2$ will dissolve. This happens because the H^+ added reacts with the OH^-, causing a decrease in the concentration of the ions on the product side, and causing the equilibrium to shift to the right.

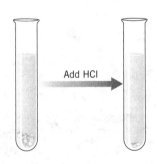

The Reaction Quotient

The reaction quotient, Q, is essentially the quantitative application of Le Châtelier's Principle.

When to Use Q

Q is calculated exactly like K, with one very important distinction: the concentrations and/or pressures used need not be the values at equilibrium. We can use values from any point in the reaction.

Q vs. K

We can compare a calculated value of Q to a known value of K to determine in which direction a net reaction will be observed as the system attempts to reach equilibrium.

Comparing Q and K	
$Q < K$	$Q > K$
If Q is less than K, then more products are needed for equilibrium. A net forward reaction will be observed.	If Q is greater than K, then less products are needed for equilibrium. A net reverse reaction will be observed.

Our thought process would go something like this:

1. I know the reaction $2\ NO_2(g) \rightleftharpoons N_2O_4(g)$ has $K_{eq} = 8$

2. The current concentrations in my flask are:

$[NO_2] = 0.100\ M$ and $[N_2O_4] = 0.025\ M$

3. The Q is currently:

$$Q = \frac{[N_2O_4]}{[NO_2]^2}$$

$$= \frac{(0.025)}{(0.100)^2}$$

$$= 2.5$$

4.

Q is less than K, so the reaction will make more products.

5.

A net forward reaction will be observed! Q will increase as the concentration of products increases and the concentration of reactants decreases.

6.

Equilibrium will be achieved, and no further reaction observed, when the value of Q is equal 8.

ICE Charts

You first learned about ICE charts in Chapter 3. However, ICE charts can also be used in equilibrium problems. In this case, ICE stands for Initial, Change, Equilibrium. An ICE chart helps us solve various kinds of equilibrium problems. Here's an example:

A flask contains only **Initial conditions**: 0.050 M H_2 and 0.050 M I_2. The value of K_{eq} for the reaction $H_2(g) + I_2(g) \rightleftharpoons 2\ HI(g)$ is 25. What will be the *equilibrium concentration* of HI be?

	$H_2(g)$ +	$I_2(g)$ $\rightleftharpoons$	2 $HI(g)$
Initial	0.050	0.050	0
Change	$-x$	$-x$	$+2x$
Equilibrium	$0.050 - x$	$0.050 - x$	$2x$

Coefficient from balanced reaction appears in change line

Equilibrium row is the sum of the Initial and the Change rows

To get from the initial conditions to equilibrium, reactants are consumed (−) and products are formed (+)

With the information from the ICE chart, we can now use the information about K_{eq}:

$$K_{eq} = 25 = \frac{[HI]^2}{[H_2][I_2]} = \frac{(2x)^2}{(0.050 - x)(0.050 - x)} = \frac{(2x)^2}{(0.050 - x)^2}$$

Take the square root of both sides:

$$5 = \frac{2x}{0.050 - x}$$

With some algebra, we find that $x = 0.036$. Therefore, the equilibrium concentration of HI($2x$) is 0.072 M.

 Ice isn't just in your summertime lemonade, it's also in AP Chemistry!

Small x Approximation

There's one trick to make the algebra involved in solving an ICE chart a little easier. Consider the ICE chart for the following reaction, which has a K_{eq} of 7×10^{-4}:

	HF(aq)	$\rightleftharpoons$	H$^+$(aq)	+	F$^-$(aq)
Initial	0.50		0		0
Change	$-x$		$+x$		$+x$
Equilibrium	$0.50 - x$		x		x

x will be very small compared to 0.50 M, so the equilibrium concentration of HF(aq) will be very close to 0.50 M, and we can simply use the initial concentration in the K_{eq} expression.

This approximation can be used when K_{eq} is $< 1.0 \times 10^{-3}$, and it will be particularly useful for problems involving weak acids (see Chapter 7).

Solubility Product Constant, K_{sp}

Roughly speaking, a salt can be considered "soluble" if more than 1 gram of the salt can be dissolved in 100 mL of water. Soluble salts are usually assumed to dissociate completely in aqueous solution. Most, but not all, solids become more soluble in a liquid as the temperature is increased.

Soluble salt

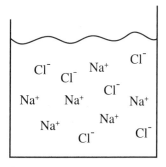

Slightly soluble or insoluble

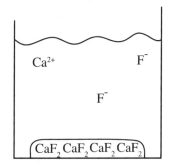

Salts that are "slightly soluble" and "insoluble" still dissociate in solution to some extent. The solubility product (K_{sp}) is a measure of the extent of a salt's dissociation in solution. The K_{sp} is one of the forms of the equilibrium expression. The greater the value of the solubility product for a salt, the more soluble the salt.

As the reactant in a K_{sp} expression is always a solid, there is never a denominator! 💬

Solubility Product

For the reaction
$$A_aB_b(s) \rightleftharpoons a\,A^{b+}(aq) + b\,B^{a-}(aq)$$

Solid reactant does not appear in K_{sp}, so no denominator

Ions appear in solubility product

Coefficients in balanced reaction become exponents

The solubility expression is $K_{sp} = [A^{b+}]^a[B^{a-}]^b$

Beaker 1: Pb^{2+}, F^-, F^-, PbF$_2$ PbF$_2$ PbF$_2$
$K_{sp} = [Pb^{2+}][F^-]^2 = 3.3 \times 10^{-8}$

Beaker 2: Pb^{2+}, Cl^-, Cl^-, Cl^-, Pb^{2+}, Cl^-, PbCl$_2$ PbCl$_2$ PbCl$_2$
$K_{sp} = [Pb^{2+}][Cl^-]^2 = 1.7 \times 10^{-5}$

K_{sp} vs. Molar Solubility

The solubility of salts can be described by the K_{sp} or by the molar solubility. The molar solubility of the salt describes the number of moles of salt that can be dissolved per liter of solution, and has units of M. (K_{sp}, like all other equilibrium constants, has no units! No shoes, no shirt, no units, no problem!)

The **molar solubility** of the salt will also be equal to the concentration of any ion that occurs in a 1:1 ratio with the salt. Typically, the molar solubility of most salts will increase with rising temperatures. This is because a higher temperature, more energy available to force the water molecules apart and make room for the solute ions.

Ion vs. Mass Solid During Evaporation

As liquid evaporates from a saturated solution, the concentration of the dissolved salt will remain constant. However, since the volume of the solution is decreasing, the number of ions in solution must decrease. These excess ions will precipitate out to form more of the solid salt.

Feast your eyes on this graph that showcases ion concentrations and mass of solid during evaporation.

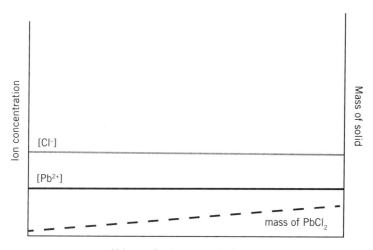

Equilibrium and the Solubility Product Constant

Solving K_{sp} Problems ❗

Don't be intimidated by K_{sp} problems—they are simply equilibrium problems. You will be able to solve many of them using the ICE chart techniques discussed earlier. You may be asked to: 1) calculate molar solubility, given the K_{sp} of a salt; 2) calculate the mass of a salt that can dissolve in a given volume of water; or 3) calculate the volume of water required to dissolve a given mass of salt. You got this.

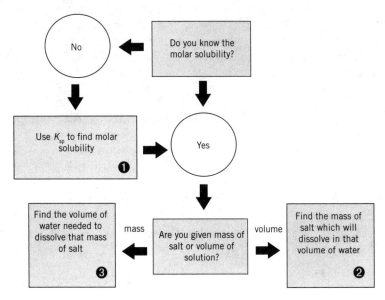

Solving K_{sp} Problems Given K_{sp}

What we are doing here is really calculating molar solubility from K_{sp}. We can determine the molar solubility of $PbCl_2$ ($K_{sp} = 1.7 \times 10^{-5}$) using the following calculations:

❶

$K_{sp} = [Pb^{2+}][Cl^-]^2$

$1.7 \times 10^{-5} = (x)(2x)^2$
$1.7 \times 10^{-5} = 4x^3$
$x = [Pb^{2+}] = 1.6 \times 10^{-2} \, M$
$[Cl^-] = 3.2 \times 10^{-4} \, M$

> Note that the concentration of chloride ions is both doubled (there will be twice the number of lead ions in solution) and squared in the K_{sp} expression. The coefficient 2 is represented twice!

Mass That Can Dissolve

Let's try calculating the mass of a salt that can dissolve in a given volume of water.

What mass of $PbCl_2$ will dissolve in 4.0 L of water?

Molar solubility of $PbCl_2 = 1.6 \times 10^{-2}$ M

Moles of $PbCl_2$ which will dissolve → $n = M \times V$ →
$(1.6 \times 10^{-2}$ $M)(4.0$ L$) = 6.4 \times 10^{-2}$ mol

Mass of $PbCl_2$ which will dissolve → $m = n \times MM$ →
$(6.4 \times 10^{-2}$ mol$)(278.1$ g/mol$) = 18$ g

Volume of Water Needed to Dissolve

Now let's try calculating the volume of a solution containing a given mass of salt.

What volume of water is needed to dissolve 5.0 g of $PbCl_2$?

Molar solubility of $PbCl_2 = 1.6 \times 10^{-2}$ M

Moles of $PbCl_2$ in 5.0 g → $n = m/MM$ →
$(5.0$ g$)/(278.1$ g/mol$) = 0.018$ mol

Volume of a saturated solution of $PbCl_2$ containing 0.018 mol $PbCl_2 =$
$V = n/M$ → $(0.018$ mol$)/(1.6 \times 10^{-2}$ $M) = 1.1$ L

Common Ion Effect

The common ion effect states that the solubility of a salt will decrease when a common ion is present.

Let's take a look at AgCl. $AgCl(s) \rightleftharpoons Ag^+(aq) + Cl^-(aq)$,
$$K_{sp} = [Ag^+][Cl^-] = 1.6 \times 10^{-10}$$

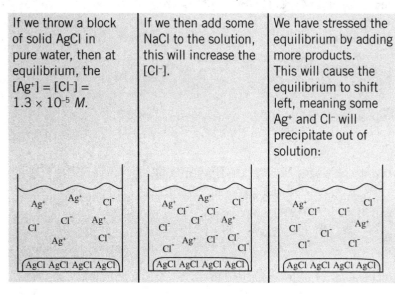

Qualitative Reasoning with Respect to pH

The same reasoning can also be applied to salts whose solubility changes with pH. An example of a salt whose solubility is affected by pH is zinc hydroxide, $Zn(OH)_2$.

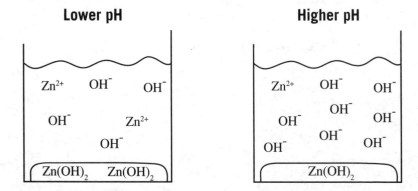

At the higher pH, more OH⁻ ions are in solution. This means that a common ion is present, and that less of the $Zn(OH)_2$ will dissolve.

Q and K_{sp}

We can use the ideas of Q and K to help us predict whether a precipitate will form.

Using Q to Predict Precipitation	
$Q < K_{sp}$	$Q > K_{sp}$
A net **forward** reaction will be observed; no precipitation will be observed.	A net **reverse** reaction will be observed; this means precipitation will occur.
$AgCl(s) \rightleftharpoons Ag^+(aq) + Cl^-(aq)$ $K_{sp} = 1.6 \times 10^{-10}$	$AgCl(s) \rightleftharpoons Ag^+(aq) + Cl^-(aq)$ $K_{sp} = 1.6 \times 10^{-10}$
If $[Ag^+] = 5.0 \times 10^{-5}$ and $[Cl^-] = 3.0 \times 10^{-7}$, then $Q = 1.5 \times 10^{-11}$.	If $[Ag^+] = 2.0 \times 10^{-3}$ and $[Cl^-] = 5.0 \times 10^{-6}$, then $Q = 1.0 \times 10^{-8}$.
This is smaller than K_{sp}, so no precipitate will be observed.	This is larger than K_{sp}, so a precipitate will form.

We can also use Q to understand the common ion effect in a quantitative way. Let's revisit the example at the beginning of this section:

If we throw a block of solid AgCl in one liter of pure water, then at equilibrium, the $[Ag^+] = [Cl^-] = 1.3 \times 10^{-5}\ M$.	If we then add 0.1 mol of NaCl to the solution, this will increase the $[Cl^-]$.	Q is greater than K_{sp}, so precipitation will be observed.
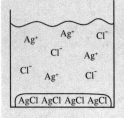	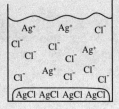	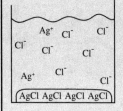
At this point, $Q = K$.	$[Ag^+]$ should still be $1.3 \times 10^{-5}\ M$, but $[Cl^-]$ is approximately $0.1\ M$. $Q = [Ag^+][Cl^-] =$ $(1.3 \times 10^{-5})(0.1) =$ 1.3×10^{-6}	At equilibrium, the concentration of $[Cl^-]$ will still be close to $0.1\ M$, so we can calculate the new equilibrium concentration of Ag^+ using K_{sp}: $[Ag^+] = K_{sp}/[Cl^-]$ $= (1.6 \times 10^{-10})/(0.1)$ $= 1.6 \times 10^{-9}$ This is much smaller than the $1.3 \times 10^{-5}\ M$ which we had in the beaker to start with!

As we've seen, reactions have a natural tendency to move to their equilibrium position. We'll explore how we can use this tendency to use reactions for energy (Chapter 6) and how these concepts are important in acid-base chemistry (Chapter 7). If you want to give yourself a little break, now would be a good time to take a walk, dance to your favorite song, or grab a snack before we move on. We'll be here when you get back!

CHAPTER 6
Redox Reactions and Electrochemistry

In this chapter we will go over processes that occur in electrochemical cells, how to calculate cell EMF, and predict spontaneity based on qualitative comparisons of cell concentrations and quantitative equations involving cell EMF, Faraday's constant, time, amperage, and moles of substances.

Oxidation States 🛈

The oxidation state (or oxidation number) of an atom indicates the number of electrons that it gains or loses when it forms a bond. For instance, upon forming a bond with another atom, oxygen generally gains two negatively charged electrons, so the oxidation state of oxygen in a bond is –2.

General Rules:

- The oxidation state of an atom that is not bonded to another atom is zero.
- The oxidation numbers for all atoms in a molecule must add up to zero.
- The oxidation numbers for all atoms in a polyatomic ion must add up to the charge on the ion.

Element/Ion/Compound	Oxidation State	Example
Neutral atom	**0** only one kind of atom present, no charge	• In Cu(s), the oxidation state on copper is zero. • In $O_2(g)$, the oxidation state on both oxygen atoms is zero. $Cu(s)^0 \quad O_2(g)^0$
Ion	**= the charge on the ion** Any ion has an oxidation state equal to the charge on that ion. This includes ions bonded to other ions in any kind of ionic compound.	• In $Zn^{2+}(aq)$ zinc has an oxidation state of +2. • In $FeCl_3$, iron is +3 and chlorine is –1. $Zn(aq)^{2+} \quad Fe^{3+} \quad Cl^-$
Oxygen	**–1 in peroxides O_2^-** **–2 in all other compounds O_2^-** In most compounds, oxygen is –2. An exception worth knowing is that in hydrogen peroxide (H_2O_2) oxygen is –1.	• In $C_6H_{12}O_6$, oxygen is –2. O^{-2} • In H_2O_2, oxygen is –1. O^{-1}

Hydrogen	**+1 when hydrogen bonded to a nonmetal**	• In CH_4, hydrogen is +1. H^{+1}
	−1 when hydrogen bonded to a metal	• In NaH, hydrogen is −1. H^{-1}
Other elements	**+1 = Group 1A Alkali Metals (Li, Na,...)** **+2 = Group 2A Alkaline Earth Metals (Be, Mg, ...)** **+3 = Group 3A (B, Al, ...)** **−2 = Group 6A (O, S,...)** **−1 = Halogens (F, Cl,...)** In the absence of oxygen, the most electronegative element in a compound will take on an oxidation state equal to its most common charge.	• In CF_4, fluorine is −1. • In CS_2, sulfur is −2.
Neutral compounds	**0** The combined oxidation states on all elements in a neutral compound must add up to zero.	In H_2SO_4 H^{+1} S^{+6} O^{-2} $(+2) + (+6) + (-8) = 0$
Ionic compounds	**+/− charge** The combined oxidation states on all elements in a polyatomic ion must add up to the charge on that ion.	In PO_4^{3-} P^{+5} O^{-2} $(+5) + (-8) = -3$

Redox Reactions

In an **oxidation-reduction** (or **redox,** for short) reaction, electrons are exchanged by the reactants, and the oxidation states of some of the reactants are changed over the course of the reaction.

Redox Reactions and Electrochemistry

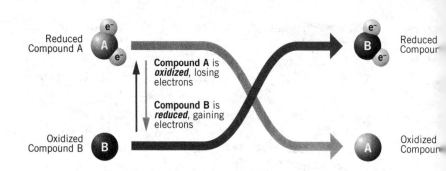

Oxidation	Reduction
1. **Atom loses valence electron(s)**	1. **Atom gains valence electron(s)**
2. **Oxidation number increases** When an atom loses a negatively charged electron, its oxidation number increases, and it is said to have been oxidized.	2. **Oxidation number decreases** When an atom gains a negatively charged electron, its oxidation number decreases, and it is said to have been reduced.
3. **Oxidation gains oxygen, loses hydrogen** "**LEO** the Lion says **GER**" LEO – **L**oss of **E**lectron(s) is **O**xidation "**OIL RIG**" OIL – **O**xidation **I**s **L**oss of electron(s)	3. **Reduction loses oxygen, gains hydrogen** GER – **G**ain of **E**lectron(s) is **R**eduction RIG – **R**eduction **I**s **G**ain of electron(s)

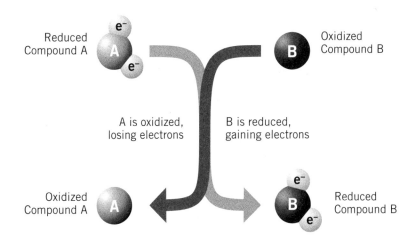

Oxidation and reduction go hand in hand. If one atom is losing electrons, another atom must be gaining them. Nothing just disappears—it's almost like a chemistry version of Newton's Third Law of Motion.

Half-Reactions

An oxidation-reduction reaction can be written as two **half-reactions**: one for the reduction and one for the oxidation. Let's look at a specific reaction as an example. Here's one.

$$Fe + 2\ HCl \rightarrow FeCl_2 + H_2$$

can be written as

$$Fe \rightarrow Fe^{2+} + 2e^- \qquad \text{Oxidation}$$
$$2\ H^+ + 2e^- \rightarrow H_2 \qquad \text{Reduction}$$

$$2\ Mg(s) + O_2(g) \rightarrow 2\ MgO(s)$$

$$2\ Mg\ +\ O_2 \rightarrow 2[Mg^{2+}\quad O^{2-}]$$

Oxidation: $Mg \rightarrow Mg^{2+} + 2e$

Reduction: $O_2 + 2e \rightarrow O^{2-}$

Standard Reduction Potentials (and Oxidation Potentials)

Every half-reaction has an electric potential, or voltage, associated with it. On the AP Chemistry Exam, you will be given the necessary values for the standard reduction potential of half-reactions for any question in which they are required. Potentials are always given as reduction half-reactions, but you can read them in reverse and flip the sign on the voltage to get oxidation potentials.

Half Reaction | Potential

F_2	+	$2e^-$	⇌	$2F^-$		+2.87 V
Pb^{4+}	+	$2e^-$	⇌	Pb^{2+}		+1.67 V
Cl_2	+	$2e^-$	⇌	$2Cl^-$		+1.36 V
Ag^+	+	$2e^-$	⇌	Ag		+0.80 V
Fe^{3+}	+	$2e^-$	⇌	Fe^{2+}		+0.77 V
Cu^{2+}	+	$2e^-$	⇌	Cu		+0.34 V
$2H^+$	+	$2e^-$	⇌	H_2		0.00 V
Fe^{3+}	+	$2e^-$	⇌	Fe		–0.04 V
Pb^{2+}	+	$2e^-$	⇌	Pb		–0.13 V
Fe^{2+}	+	$2e^-$	⇌	Fe		–0.44 V
Zn^{2+}	+	$2e^-$	⇌	Zn		–0.76 V
Al^{3+}	+	$2e^-$	⇌	Al		–1.66 V
Mg^{2+}	+	$2e^-$	⇌	Mg		–2.36 V
Li^+	+	$2e^-$	⇌	Li		–3.05 V

(left axis: increasing strength as an oxidizing agent ↑; right axis: increasing strength as a reducing agent ↓)

Look at the reduction potential for Zn^{2+}.

$Zn^{2+} + 2e^- \rightarrow Zn(s)$ $\qquad E° = -0.76$ V

Read the reduction half-reaction in reverse and change the sign on the voltage to get the oxidation potential for Zn.

$Zn(s) \rightarrow Zn^{2+} + 2e^-$ $\qquad E° = +0.76$ V

The larger the potential for a half-reaction, the more likely it is to occur.

$F_2(g) + 2e^- \rightarrow 2F^-$ $\qquad E° = +2.87$ V

$F_2(g)$ has a very large reduction potential, so it is likely to gain electrons and be reduced.

Li(s) → Li⁺ + e⁻ $E° = +3.05$ V

Li(s) has a very large oxidation potential, making it very likely to lose electrons and be oxidized. You can calculate the potential of a redox reaction if you know the potentials for the two half-reactions that constitute it.

There are two important things to remember when calculating the potential of a redox reaction:

- Add the potential for the oxidation half-reaction to the potential for the reduction half-reaction.
- Never multiply the potential for a half-reaction by a coefficient.

Let's look at the following reaction:

$$Zn + 2\ Ag^+ \rightarrow Zn^{2+} + 2\ Ag(s)$$

The two half-reactions are:

Oxidation: $Zn \rightarrow Zn^{2+} + 2e^-$ $E° = +0.76$ V

Reduction: $Ag^+ + e^- \rightarrow Ag(s)$ $E° = +0.80$ V

$$E = E_{oxidation} + E_{reduction}$$

$$E = 0.76\ V + 0.80\ V = 1.56\ V$$

Notice that we ignored that silver has a coefficient of 2 in the balanced equation.

The relative reduction strengths of two different metals can also be determined qualitatively. In the above reaction, if Zn(s) were placed in a solution containing Ag⁺ ions, the silver ions have a high enough reduction potential that they would take electrons from the zinc and start forming solid silver, which would precipitate out on the surface of the zinc.

However, if Ag(s) were placed in a solution containing Zn^{2+} ions, zinc does not have a high enough reduction potential to take electrons from silver and so no reaction would occur. So, when a solid metal is placed into a metallic solution and a new solid starts to form, the reduction potential of the metal in solution is greater than that of the solid metal. If no solid forms, the reduction potential of the solid metal is higher.

 Remember!

Balance <u>ATOMS FIRST</u> by multiplying coefficients, then balance the <u>CHARGE LAST</u> by adding electrons.

> **Final Thought:** Don't worry about reviewing the manner by which you balance redox in acidic or basic media—you won't need to know that for the AP Chemistry Exam, so don't waste any precious time on it.

Common Redox Reactions

There are several very common types of redox reactions that would be worth being familiar with on the new exam. The most common is what is typically called a "single replacement" reaction in chemistry. One example is the reaction that occurs when solid zinc is dropped into a beaker of copper nitrate:

$$Cu^{2+}(aq) + Zn(s) \rightarrow Zn^{2+}(aq) + Cu(s)$$

In the above reaction, the Cu^{2+} has a higher reduction potential than the Zn^{2+}. So, the copper ions take electrons from the zinc metal, causing a reaction to occur. Note that if a piece of solid copper were dropped into a solution containing Zn^{2+} ions, no reaction would occur!

> ⊘ The terms "reducing agent" and "oxidizing agent," while commonly used in chemistry, are outside the scope of this exam.

Redox Reactions and Electrochemistry

Another common redox reaction is the one which occurs when a metal dissolves in acid (H⁺ ions). The reduction of H⁺ into $H_2(g)$ is:

$$2\ H^+(aq) + 2e^- \rightarrow H_2(g)$$

If a solid piece of magnesium were dissolved in acid, the net ionic reaction would be:

$$Mg(s) + 2\ H^+(aq) \rightarrow Mg^{2+}(aq) + H_2(g)$$

Another type of dissolving can occur when metals with a weak hold on their electrons, such as the alkali metals, are dropped in water. If a piece of sodium were dropped in water, the water would steal its electrons, leading to the sodium dissolving via the following net ionic:

$$2\ H_2O(l) + 2\ Na(s) \rightarrow H_2(g) + 2\ Na^+(aq) + 2\ OH^-(aq)$$

Finally, there are several energy-producing processes that are redox reactions, including hydrocarbon combustions and the metabolism of sugars, fats, and proteins.

Combustion of propane: $C_3H_8 + 5\ O_2 \rightarrow 4\ H_2O + 3\ CO_2$

Metabolism of glucose: $C_6H_{12}O_6 + 6\ O_2 \rightarrow 6\ CO_2 + 6\ H_2O$

Redox Titrations

A titration involves the slow addition of a solution at a known concentration to another solution of unknown concentration in order to determine the concentration of the unknown solution.

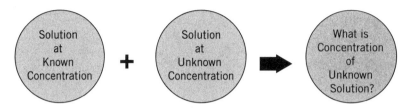

To determine the endpoint of a titration, a color change is often used. Titrations are frequently used in acid-base reactions (more on that in Chapter 7), but redox reactions can also be used in titration situations.

Potassium permanganate, $KMnO_4$, is frequently used in redox titrations. The manganese ion has an oxidation state of +7 in the permanganate ion (MnO_4^-), and a solution containing permanganate ions is a deep purple color. The manganese in MnO_4^- reduces to Mn^{2+} (thus changing its oxidation state to +2) when mixed with compounds that can be oxidized. Mn^{2+} is a colorless ion.

Frequently, when potassium permanganate is titrated into a colorless solution that contains a compound that can be oxidized, the end of the titration is marked by the solution turning pink. Initially, the permanganate ions take electrons from the oxidized compound and reduce to Mn^{2+}. However, once the compound that is being oxidized runs out, there are no electrons left for the MnO_4^- to take, and thus any extra permanganate ion that is added remains unreduced and retains its purple color, which when diluted sufficiently appears pink.

Example:

$8 H^+(aq) + MnO_4^-(aq) + 5 Fe^{2+}(aq) \rightarrow 5 Fe^{3+}(aq) + Mn^{2+}(aq) + 4 H_2O(l)$

20.0 mL of a solution containing $Fe(NO_3)_2$ is titrated with some 0.155 M $KMnO_4$, causing the above reaction to occur. If the endpoint is achieved after 25.13 mL of $KMnO_4$ is added, what is the initial concentration of the $Fe(NO_3)_2$?

The NO_3^- and K^+ ions are spectators, and so they can be safely ignored. First, we need the moles of MnO_4^- added to get to the endpoint:

$$0.155 \ M \ MnO_4^- = \frac{n}{0.02513 \ L} \quad n = 0.00390 \ mol \ MnO_4^-$$

Then, we can use the coefficients in the balanced equation to convert that to moles of Fe^{2+}:

$$0.00390 \ mol \ MnO_4^- \times \frac{5 \ mol \ Fe^{2+}}{1 \ mol \ MnO_4^-} = 0.0195 \ mol \ Fe^{2+}$$

Finally, we can divide by the initial volume of Fe^{2+} to arrive at the answer.

$$0.0195 \ mol \ Fe^{2+} = 0.973 \ M = [Fe^{2+}] = [Fe(NO_3)_2]$$

Electrochemistry

Units

To study electrochemistry, we must first gain an understanding of the language. Electrochemistry is all about the movement of electrons, which carry charge. Charge is measured in Coulombs (C), and one mole of electrons carries 96,500 C of charge. This is called Faraday's constant, symbolized by the letter F.

Current is the flow of electrons over time:

$$I = q/t$$

I is current (measured in Amperes).
q is charge (measured in Coluombs).
t is time (measured in seconds).

Electric potential, aka voltage (V), is defined as the amount of energy (in Joules) carried per unit of charge (in Coulombs).

Voltaic Cells

Did you know that all batteries, including the one in your cell phone, depend on chemistry to work? All batteries are voltaic cells (aka galvanic cells), which are cells in which electrons flow from a location of greater potential energy, called the anode, to one of less potential energy, called the cathode.

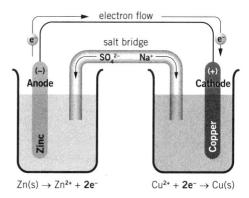

In the diagram above, the zinc electrode is the anode, and the copper electrode is the cathode. Oxidation always occurs at the anode, and the lost electrons carry with them the potential energy as they flow to the cathode, where the reduction always occurs.

⊘ We are absoluty positive that you do not need to be able to label electrodes as negative or positive!

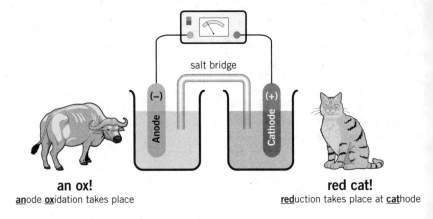

an ox!
<u>an</u>ode <u>ox</u>idation takes place

red cat!
<u>red</u>uction takes place at <u>cat</u>hode

As the reaction progresses, the anode will gradually get more positive as Zn^{2+} ions are added to the solution. The cathode, on the other hand, will get more negative, as Cu^{2+} ions are removed from the solution. That creates a problem—electrons will not flow from a positive (zinc) electrode to a negative (copper) one.

To fix this problem, a salt bridge is used. The salt bridge always contains a soluble salt (Na_2SO_4 in this case). As the reaction progresses, the SO_4^{2-} anions are attracted to the anode and the Na^+ cations are attracted to the cathode. This makes both electrodes stay neutral and allows the reaction to continue. Without the salt bridge, the cell potential would drop to 0!

Note that electrons do NOT flow through the salt bridge! Instead, they flow through the wire connecting the electrodes. This is an important distinction!

Cell Potential

To determine the cell potential ($E°_{cell}$) of a voltaic cell, we have to know the standard reduction potentials (SRPs) for both the anode and the cathode. An SRP is a measurement of how likely something is to be reduced. SRPs can be positive or negative. To determine $E°_{cell}$, simply subtract the reduction potential at the anode from the reduction potential at the cathode.

$$Cu^{2+}(aq) + Zn(s) \rightarrow Zn^{2+}(aq) + Cu(s)$$

Example: Given the following SRPs, determine the cell potential for a zinc/copper voltaic cell.

Half-Reaction	Standard Reduction Potential (V)
$Cu^{2+}(aq) + 2e^- \rightarrow Cu(s)$	0.34
$Zn^{2+}(aq) + 2e^- \rightarrow Zn(s)$	–0.76

The Cu^{2+} is reduced at the cathode, and the $Zn(s)$ is oxidized at the anode: 0.34 V – (–0.76 V) = 1.10 V.

A VERY common mistake students make when calculating $E°_{cell}$ is to apply the coefficients that appear in a balanced equation to the cell potentials. Don't do this!

$$2\ Ag^+(aq) + Pb(s) \rightarrow Pb^{2+}(aq) + 2\ Ag(s)$$

Example: Given the following SRPs, determine the cell potential for the silver/lead voltaic cell.

Half-Reaction	Standard Reduction Potential (V)
$Ag^+ + e^- \rightarrow Ag(s)$	0.80
$Pb^{2+} + 2e^- \rightarrow Pb(s)$	–0.13

The Ag^{2+} is reduced at the cathode, and the $Pb(s)$ is oxidized at the anode. The "2" coefficient in front of the Ag is irrelevant for our purposes: 0.80 V – (–0.13 V) = 0.93 V.

You will be given any SRPs you need on the test, so don't worry about memorizing them!

Note that in both examples above, the cell potential is positive. That will always be the case with galvanic cells. In fact, a positive cell potential is another way to determine favorability. As you may recalled, ΔG is another measurement of favorability. The two concepts are related via the following equation:

$$\Delta G = -nF\mathcal{E}$$

ΔG is Gibbs free energy (in Joules)
n is the number of moles of electrons transferred (in mol e^-)
F is Faraday's constant (96,500 C/mol e^-)
$\mathcal{E}$ is cell potential (in V, aka J/C)

Example: What would the Gibbs free energy value be for the silver/lead voltaic cell from the previous example?

The number of moles of electrons can be determined by the coefficient on the electrons in the balanced equation before they canceled out. In this case, that's 2.

$\Delta G = -(2 \text{ mol } e^-)(96{,}500 \text{ C/mol } e^-)(0.93 \text{ J/C})$

$\Delta G = -180{,}000$ J or -180 kJ

Note that the negative sign in the equation means that when $\mathcal{E}$ is positive, ΔG will always be negative.

The following illustrates the relationship between favorability: the standard free energy change $\Delta G°$, cell potential $E°$, and the equilibrium constant K. Remember, everything in the universe depends on available free energy $\Delta G°$. Electrochemistry is like a triangle of related concepts, or almost like a puzzle and you don't see the big picture until you fit in all of the pieces.

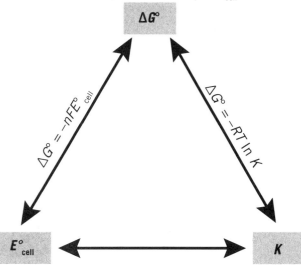

Interrelationship Between $\Delta G_r°$, $E°_{cell}$, and K

$\Delta G°$	K	$E°_{cell}$	Reaction at Standard-state Conditions
< 0	> 1	> 0	Favored
0	1	0	At equilibrium
> 0	< 1	< 0	Not favored

Redox Reactions and Electrochemistry

Non-Standard Conditions

It's time to talk more about the "Standard" part of "Standard Reduction Potential." In electrochemistry, standard means solution concentrations of 1.0 M, a temperature of 25°C, and a partial pressure of 1.0 atm for any gases.

You might ask, what happens if the conditions are NOT standard? For the purposes of the AP Exam, you won't have to calculate that. However, you should be able to determine whether a cell potential will be greater than, less than, or the same as standard conditions if the conditions change.

The way we do that is by using the reaction quotient. When concentration or pressure changes, the value for the reaction quotient will change.

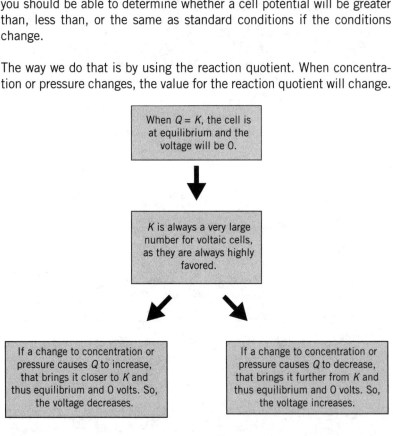

Please note that voltaic cells are NOT in equilibrium, and thus, you cannot use Le Châtelier's Principle to discuss changes in cell potential due to deviations from standard conditions. **You must use Q vs. K!**

Let's take a look at how this would work in practice.

$$Cu^{2+}(aq) + Zn(s) \rightarrow Zn^{2+}(aq) + Cu(s)$$

Example: The zinc/copper galvanic cell has a voltage of 1.10 V at standard conditions. If the concentration of the Zn^{2+} were increased to 2.0 M while the concentration of Cu^{2+} was held constant at 1.0 M, how would that affect the cell potential?

$$Q = \frac{\left[Zn^{2+}\right]}{\left[Cu^{2+}\right]}$$

If $[Zn^{2+}]$ increases while $[Cu^{2+}]$ does not change, Q will increase. This brings the reaction closer to equilibrium (and zero volts), so the overall potential will decrease.

The above example tells us how changes from standard conditions would affect cell potential. One other question that might be asked is how changes to cell potential might affect battery life. If the concentration of both Zn^{2+} and Cu^{2+} the above example were initially 0.50 M, that would not change Q, and so the voltage would not change.

However, because there is a much lower concentration of both species, the battery would last half as long. Simply put, the greater the concentration of each species, the longer the battery will last, independent of any changes to the cell potential.

This is also true for the electrodes—if either the copper or zinc electrode were replaced with a greater mass, the battery would last longer. However, that would have no affect on the voltage, as solids do not appear in the equilibrium expression.

⊘ The effects of temperature on cell potential and the Nernst equation are excluded from the AP Chemistry Exam.

Electrochemistry Flow Chart

Electrochemistry is dynamic and there's so much going on at the microscopic level—the electrons are doing all of the work! There's electrons, there's ions, there's all of these moving parts in solid, liquid, and aqueous phases, and there is thermodynamics and equilibrium tied in. For the AP Chemistry Exam, be sure that you understand the "Big Picture" before trying to figure out all of the little pieces. So let's review the big picture now.

For the AP Chemistry Exam, you will be expected to draw out at the particle level:

1. Flow of electrons in the wire
2. Flow of ions in the electrolyte solution
3. Gain or Loss of mass at the cathode/anode.

A common galvanic cell is the Daniell cell, and it highlights those three areas because it highlights what the AP Chem Exam will likely ask you to drawn—electrons moving, ions flowing, and mass gained or lost. Electrons flow from the anode to the cathode through an external wire. Check out this Daniell cell, shown below.

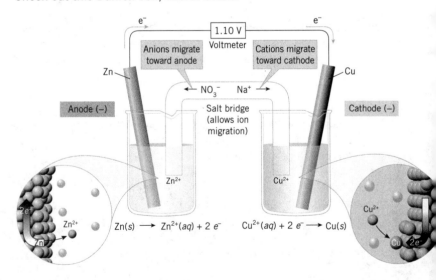

The Zn(s) gives up its electrons to form Zn^{2+}(aq) ions. The electrons remain behind on the Zn electrode. Since Zn is oxidized, the Zn electrode is the anode.

The electrons travel through an external circuit to the copper electrode. Here the Cu^{2+}(aq) ions in contact with the Cu electrode accept these electrons and become Cu(s). Since Cu^{2+} is reduced, the Cu electrode is the cathode.

So, in a galvanic cell, electrons flow from anode to cathode through an external circuit.

Electrolytic Cells

Hydrolysis

Not all electrochemical cells proceed without the input of outside energy. A cell that needs an outside source of current (such as from a battery) to occur is called an electrolytic cell.

One of the most common electrolytic reactions is the electrolysis of water. If you run current through a solution of pure water, the water will be both oxidized and reduced, with the net ionic equation being $2\ H_2O(l) \rightarrow 2\ H_2(g) + O_2(g)$.

The hydrogen is reduced while the oxygen is oxidized. If the gases were to be collected, as in the diagram below, you would expect to see twice as much hydrogen due to the "2" coefficient in the balanced equation.

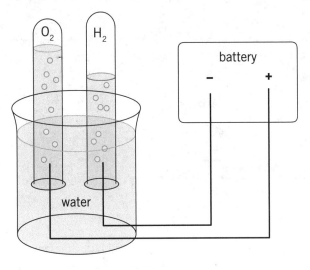

If you were to run current through an ionic solution, the reaction that occurs would depend on the relative SRPs of the ions in solution versus water. At the cathode, either water or the cation would be reduced, depending on which one has the higher SRP. At the anode, either water or the anion would be oxidized, but in this case, it's whichever has the LOWER SRP.

Example: A current is run through a solution of nickel (II) chloride. Using the SRPs given below, identify the net ionic reaction which would occur.

Half-Reaction	Standard Reduction Potential
$Cl_2(g) + 2e^- \rightarrow 2\ Cl^-$	1.36 V
$O_2(g) + 4\ H^+(aq) + 4e^- \rightarrow 2\ H_2O(l)$	1.23 V
$Ni^{2+}(aq) + 2e^- \rightarrow Ni(s)$	–0.25 V
$2\ H_2O(l) + 2e^- \rightarrow H_2(g) + 2\ OH^-(aq)$	–0.83 V

We'll look at the cathode first. The two options for reduction are water (the bottom reaction shows the reduction of water) or the Ni^{2+}. Between the two, the Ni^{2+} has the higher SRP, and so it will be reduced at the cathode.

At the anode, we have to flip all the SRPs to give oxidation reactions. When doing so, the two options for being oxidized are either the water (the second from the top reaction would be the oxidation of water when flipped) or the chloride. Between the two, the water has the LOWEST SRP, so that's the oxidation that would occur.

Reduction: $Ni^{2+}(aq) + 2e^- \rightarrow Ni(s)$

Oxidation: $2\ H_2O(l) \rightarrow O_2(g) + 4\ H^+(aq) + 4e^-$

Net: $2\ H_2O(l) + 2\ Ni^{2+}(aq) \rightarrow 2\ Ni(s) + O_2(g) + 4\ H^+(aq)$

Electroplating

If a current is run through a solution containing a metallic cation, such as in the example above, that cathode can "plate" out into a metal at the cathode. The mass of the metal that is collected can be calculated via stoichiometry.

Example: A current of 0.500 A is run through a solution of $CuCl_2(aq)$ for 3.00 minutes. What mass of copper metal will plate out at the cathode?

First, we have to determine the charge applied, keeping in mind that an amp is a Couloumb per second:

$$\frac{0.500\,C}{1.0\,s} \times 180\,s = 90.0\,C$$

Then, we use Faraday's constant to convert from Coulombs to moles of electrons. After that, we know that the copper ion has a charge of 2+, so it requires 2 moles of electrons to create one mole of copper. Finally, we convert moles of copper to grams for our final answer.

$$90.0\,C \times \frac{1\ mol\ e^-}{96500\,C} \times \frac{1\ mol\ Cu}{2\ mol\ e^-} \times \frac{63.55\ g\ Cu}{1\ mol\ Cu} = 0.0296\ g\ Cu$$

CHAPTER 7
Acids and Bases

Arguably the single most covered topic on the AP Exam is acids and bases. There's a lot that goes into this particular topic, but it's best to start at the beginning: with definitions. What is an acid? Conversely, what is a base?

Definitions !

Arrhenius !

An Arrhenius acid is a substance that dissociates to produce H⁺ ions

Acids and Bases

Hydrochloric Acid

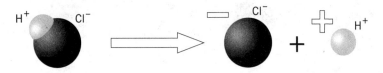

An Arrhenius base is a substance that dissociates to produce OH– ions.

Sodium Hydroxide

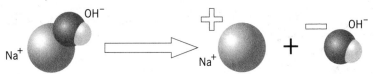

The problem with the Arrhenius definition is that while all bases do cause the creating of OH⁻ in solution, they do not always do so directly.

A Brønsted-Lowry acid is a proton (H+) donor. A Brønsted-Lowry base is a proton acceptor.

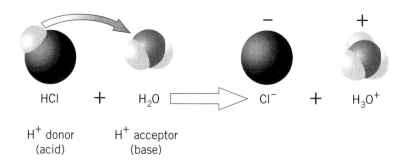

Substances such as water which can act as either an acid or a base are called amphoteric. A good way to remember this is that "amphi" means "two" or "both sides," like the word "amphibious," which is something that can live on land or water, just as water of substances that are amphoteric can act as either acid or base. It can be on both sides of the issue!

Two substances that are separated from each other by a single proton are called a conjugate pair.

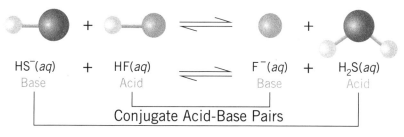

Acids and Bases

pH 🛑

"p" in acid/base chemistry simply means $-\log_{10}$.

$$pH = -\log[H^+]$$

The higher the concentration of H^+ in solution, the lower the pH is.

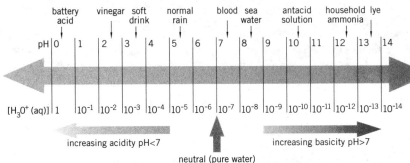

The pH Scale

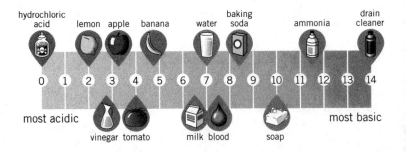

Another less common term you may see use is pOH.

$$pOH = -\log[OH^-]$$

K_w

As stated, water can act as either an acid or a base. This is true even when it's mixed with itself, as shown below where the first water molecule acts as an acid by donating a proton, and the second one accepts that proton and thus acts as an base.

$$H_2O(l) + H_2O(l) \rightleftharpoons OH^- + H_3O^+(aq)$$

The H_3O^+ ion is called the hydronium ion. The above reaction could also be written as

$$H_2O(l) \rightleftharpoons OH^-(aq) + H^+(aq)$$

In either case, the equilibrium constant for that expression, called K_w, would be

$$K_w = [H^+][OH^-]$$

For pure water at 25°C, K_w is always equal to 1.0×10^{-14}.

If you rearrange the K_w expression, you can also get pH + pOH = 14.

An aqueous solution has a pH of 5.4 at 25°C. What is the [OH$^-$] present in the solution?

There are two approaches to solving a problem like this.

Option 1:

5.4 + pOH = 14
pOH = 8.6
$-\log[OH^-] = 8.6$
$10^{-8.6} = 2.5 \times 10^{-9}$ M

Option 2:

$-\log[H^+] = 5.4$
$10^{-5.4} = 4.0 \times 10^{-6}$ M
$1.0 \times 10^{-14} = (1.0 \times 10^{-6})[OH^-]$
$[OH^-] = 2.5 \times 10^{-9}$ M

The auto-ionization of water is an endothermic process, so any increase in temperature would cause a shift to the right.

This DOES NOT make the water more basic or acidic. As long as [H⁺] = [OH⁻], a solution is neutral.

Temperature (°C)	K_w	pK_w	pH of water
10	2.9×10^{-15}	14.54	7.27
25	1.0×10^{-14}	14.00	7.00
50	5.5×10^{-14}	13.26	6.63
100	5.1×10^{-13}	12.28	6.14

As you can see in this jazzy table, the warmer the water, the lower the pH. It may not be the "coolest" trend (heh, heh) but it's definitely a trend.

Strength 🔔

Strong Acids/Bases 🔔

An acid is considered strong if it dissociates completely in solution. A base is also considered strong if it dissociates completely in solution. Remember that as: strong by dissociation.

Strong Acids	Strong Bases
Hydrochloric Acid, HCl	Group I hydroxides (NaOH, LiOH, KOH, etc.)
Hydroiodic Acid, HI	$Sr(OH)_2$, $Ba(OH)_2$
Hydrobromic Acid, HBr	Ammonium hydroxide, NH_4OH
NitricAcid, HNO_3	
Sulfuric Acid, H_2SO_4	
Perchloric Acid, $HClO_4$	

To find the pH of any strong acid, all you need to do is take the -log of the concentration of the acid.

What is the pH of a 1.50 M solution of HBr?

$$pH = -\log[H^+]$$
$$-\log(0.75) = -0.17$$

As shown above, pH is NOT limited to the 0–14 range.

Weak Acids 🛑

Weak acids are those that do not dissociate fully in solution.

To calculate the pH of a weak acid, we need to revisit equilibrium.

$$HF(aq) \rightleftharpoons F^-(aq) + H^+(aq)$$

$$K_a = \frac{[F^-][H^+]}{[HF]}$$

What is the pH of a 1.0 M HF solution? ($K_a = 6.6 \times 10^{-4}$)

Using an ICE chart:

HF	F⁻	H⁺
1.0	0	0
$-x$	$+x$	$+x$
$1.0 - x$	x	x

Using the "small x" approximation means the "$-x$" in "$1.0 - x$" is insignificant and can be ignored.

$$6.6 \times 10^{-4} = \frac{(x)(x)}{1.0}$$

$$x = 0.026 \; M = [H^+]$$

$$-\log(0.026) = 1.59$$

Weak bases ionize (instead of dissociate), but the same general logic applies to a weak base calculations.

What is the pH of a 0.50 M NH_3 solution? ($K_b = 1.8 \times 10^{-5}$)

$$NH_3(aq) + H_2O(l) \rightleftharpoons NH_4^+(aq) + OH^-(aq)$$

$$K_b = \frac{[NH_4^+][OH^-]}{[NH_3]}$$

$$1.8 \times 10^{-5} = \frac{(x)(x)}{0.50}$$

$$x = 0.0030 \; M = [OH^-]$$

(rest of this solution is on next page)

Acids and Bases

$$\text{pOH} = -\log(0.0030)$$
$$\text{pOH} = 2.52$$
$$14 = \text{pH} + 2.52$$
$$\text{pH} = 11.48$$

A list of some of the some common weak acids and bases is below.

Weak Acids	Weak Bases
Acetic Acid	**Ammonia**
Carbonic Acid	**Methylamine**
Formic Acid	**Ethylamine**
	Pyridine

Many weak acids have a –OH group attached to a carbon that is also double bonded to a different oxygen. These are called carboxylic acids.

Many weak bases contain nitrogen. These are commonly called nitrogenous compounds.

Polyprotic Acids

Some acids have the ability to donate more than one proton to a solution. These acids are called polyprotic.

$$H_3PO_4(aq) \rightleftharpoons H^+(aq) + H_2PO_4^-(aq) \qquad K_{a1} = 7.1 \times 10^{-2}$$

$$H_2PO_4^-(aq) \rightleftharpoons H^+(aq) + HPO_4^{2-}(aq) \qquad K_{a2} = 6.3 \times 10^{-8}$$

$$HPO_4^{2-} \rightleftharpoons H^+(aq) + PO_4^{3-}(aq) \qquad K_{a3} = 4.5 \times 10^{-15}$$

Each dissociation is increasingly less likely to occur as the H^+ ions floating around in solution after the first dissociation cause a left shift.

In a solution of phosphoric acid, the species with the highest concentration would be the H_3PO_4, and that with the lowest concentration would be the PO_4^{3-}.

Percent Dissociation

Percent dissociation (or ionization) describes what percentage of a weak acid (or base) molecule will dissociate (or ionize) in solution. We'll use "HA" to symbolize a generic weak acid in the formula.

For $HA + H_2O(l) \rightleftharpoons H_3O^+(aq) + A^-(aq)$

$$\% \text{ dissociation} = \frac{[H_3O^+]}{[HA]} \times 100\%$$

Acids and Bases

If given the pH of a solution, you can often calculate the percent dissociation pretty easily.

A solution of 2.0 M HNO_2 has a pH of 2.67. What is its percent dissociation?

$$-\log[H_3O^+] = 2.67$$

$$[H_3O^+] = 0.0021\ M$$

$$\frac{0.021}{2.0 - 0.0021} \times 100\ \% = 0.11\ \%$$

As acid concentration decreases, percent ionization increases. This is because at lower concentrations, the reverse reaction (A^- reacting with H_3O^+) is less likely to occur. The forward reaction (HA with H_2O) rate does not change with concentration because there are a lot more reactant molecules in solution.

Acid Concentration ↓	% Ionization ↑

Binary Acids and Oxoacids 💀

A binary acid is one which consists of only two elements, usually hydrogen and a halogen. The more electronegative the halogen, the more effectively it "hangs on" to its proton, and the weaker the acid will be.

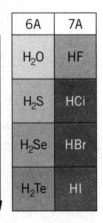

An oxoacid is H-O-X, where X is a halogen. The more electronegative the halogen is, the more it attracts the electrons in oxygen, weakening the H-O bond.

Shift of electron density

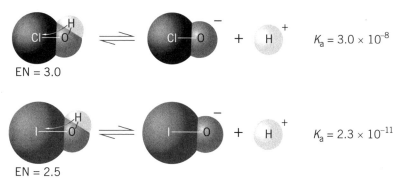

Acid/Base Reactions

Acids will transfer protons to bases in all acid/base reactions. In net ionic equations, strong acids always appear as H^+ and strong bases appear as OH^-.

Type	Reaction Example	Net Ionic Equation
Strong Acid/Strong Base	HCl and NaOH	$H^+(aq) + OH^-(aq) \rightarrow H_2O(l)$
Strong Acid/Weak Base	HCl and NH_3	$H^+(aq) + NH_3(aq) \rightarrow NH_4^+(aq)$
Weak Acid/Strong Base	HF and NaOH	$HF(aq) + OH^-(aq) \rightarrow H_2O(l) + F^-(aq)$
Weak Acid/Weak Base	HF and NH_3	$HF(aq) + NH_3(aq) \rightleftharpoons F^-(aq) + NH_4^+(aq)$

Salt Hydrolysis 🛑

If you dissolve a salt in water, the ions that make up that salt can react with water to create basic or (rarely) acidic solutions.

Most anions, with their negative charge, can attract a proton from water. Carbonate is an example:

$$CO_3^{2-} (aq) + H_2O(l) \rightleftharpoons HCO_3^- (aq) + OH^-(aq)$$

Thus, if K_2CO_3 were dissolved in water, the solution would be basic.

The only cations that CANNOT do this are the conjugates of the six strong acids, which are ineffective bases.

Non-basic anions:

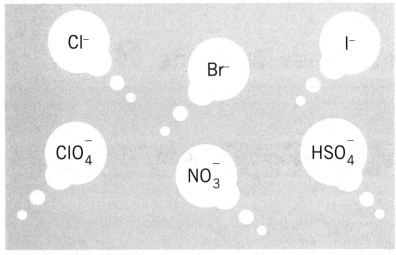

For cations to create an acidic environment, they need to be able to donate a proton. This is rare, although NH_4^+ can do it. A solution of NH_4NO_3 would be acidic via the following reaction:

$$NH_4^+ (aq) + H_2O(l) \rightleftharpoons NH_3(aq) + H_3O^+(aq)$$

If the salt contains the conjugate base of a strong acid and a cation that CANNOT donate protons, it will create a neutral solution. NaCl and MgI_2 are two examples.

Concentration Effects

When dealing with polyprotic acids that react with a strong base, the number of protons that can be removed is dependent on the concentration of the base.

0.1 M H_2CO_3 + 0.1 M NaOH: $H_2CO_3(aq) + OH^-(aq) \rightarrow$

$$HCO_3^- (aq) + H_2O(l)$$

0.1 M H_2CO_3 + 0.2 M NaOH: $H_2CO_3(aq) + 2\ OH^-(aq) \rightarrow$

$$CO_3^{2-} (aq) + 2\ H_2O(l)$$

When dealing with strong acids and a base that can accept multiple protons, the same logic applies.

0.1 M HCl + 0.1 M Na_3PO_4: $H^+(aq) + PO_4^{3-} (aq) \rightarrow HPO_4^{2-} (aq)$

0.2 M HCl + 0.1 M Na_3PO_4: $2\ H^+(aq) + PO_4^{3-} (aq) \rightarrow H_2PO_4^- (aq)$

The concentration of the strong acid or base always dictates the number of proton transfers.

pH Change 🛑

If a strong base is added to a strong acid, the pH will change easily. This is because the OH⁻ ions always react with the dissociated H⁺ ions, changing [H⁺] and thus pH.

However, if a strong base is added to a weak acid, the pH does not change as much. This is because the strong base will react primarily with the undissociated HA molecules, which has little effect on pH.

The same is true for adding a strong acid to a base; a weak base will resist change in pH more effectively than a strong base.

Buffers 🛑

A buffer is a solution that resists change in pH. Buffers consist of weak acids and their conjugate base in similar amounts.

Adding a strong acid or base just creates more of the conjugate. This keeps the pH fairly stable.

For a buffer of acetic acid ($HC_2H_3O_2$) and sodium acetate ($NaC_2H_3O_2$):

Adding a strong acid:

$$C_2H_3O_2^-(aq) + H^+(aq) \rightleftharpoons HC_2H_3O_2(aq)$$

Adding a strong base:

$$HC_2H_3O_2(aq) + OH^-(aq) \rightleftharpoons C_2H_3O_2^-(aq) + H_2O(l)$$

Strong acids cannot make buffers because the conjugates of strong acids are ineffective bases.

Buffers that consist of a weak base and its conjugate exist too, but are rare. A NH_3/NH_4Cl buffer is a common example.

Henderson-Hasselbalch 🛑

To calculates the pH of a buffer, the Henderson-Hasselbalch equation can be used.

$$pH = pK_a + \log [A^-]/[HA]$$

pK_a is $-\log K_a$

[A⁻] is the concentration of the conjugate base.
[HA] is the concentration of the weak acid.

Acids and Bases

Example:

A buffer is created by dissolving 2.50 g of NaClO into 150 mL of 0.250 M HClO (K_a = 3.5 × 10^{-8}). What is the pH of the resulting solution?

First, we have to figure out the concentration of the ClO$^-$.

$$2.50 \text{ g NaClO} \times \frac{1 \text{ mol NaClO}}{74.44 \text{ g NaClO}} = \frac{0.0336 \text{ mol NaClO}}{0.150 \text{ L}} = 0.224 \, M$$

= [NaClO] = [ClO$^-$]

Then, we plug everything in to Henderson-Hasselbalch.

pH = –log (3.5 × 10^{-8}) + log $\frac{(0.224)}{(0.250)}$
pH = 7.46 + log (0.896)
pH = 7.41

The pH of a buffer depends on the ratio of [A$^-$]:[HA].

Buffer capacity describes how effectively a buffer can resist change in pH.

More concentrated buffers have a greater buffer capacity.

A 1.0 M HC$_2$H$_3$O$_2$/1.0 M NaC$_2$H$_3$O$_2$ buffer and a 0.1 M HC$_2$H$_3$O$_2$/0.10 M NaC$_2$H$_3$O$_2$ would have the same pH, but the 1.0 M buffer would have a greater buffer capacity.

If the solution pH > pK_a of the weak acid, that means [A$^-$] > [HA]

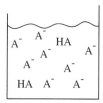

If the solution pH < pK_a of the weak acid, that means [HA] > [A$^-$]

There is a basic/pOH version of Henderson-Hasselbalch as well for weak base buffers.

$$pOH = pK_b + \log [BH^+]/[B]$$

pK_b is $-\log K_b$

[BH$^+$] is the concentration of the conjugate acid.
[B] is the concentration of the weak base.

> **Exclusion Statement**
>
> You do not know how to calculate the change in pH when an acid or base is added to a buffer.

Acids and Bases

Titrations !

An acid/base titration is a method of discovering the concentration of an unknown acid or base.

The solution of unknown concentration is usually placed in a flask.

The solution of known concentration is poured into the buret, like the one shown below.

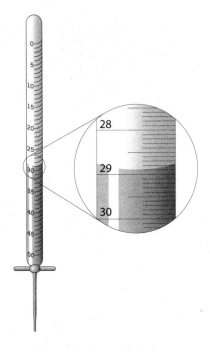

Burets are read upside-down; in the above diagram the buret reads 28.90 mL.

The solution in the buret is then titrated into the flask until the endpoint is reached.

Volume in a buret is tracked via volume change, not actual volume.

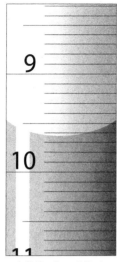
Initial Reading = 9.63 mL

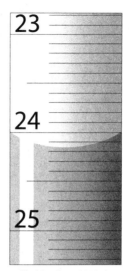
Final Reading = 24.16 mL

In the above diagram, the total volume dispensed is 24.16 mL − 9.63 mL = 14.53 mL.

The equivalence point of a titration occurs when the moles of acid = moles of base. The equivalence region is the area of rapid pH change.

30.0 mL of HCl is titrated with some 0.95 M NaOH. The endpoint is reached after 24.80 mL of NaOH is added. What is the concentration of the HCl?

At equivalence, there are (0.95 M)(0.02480 L) = 0.024 mol of base. In this example, the moles of acid and moles of base are in a 1:1 ratio, so there are 0.024 mol of acid present in the flask.

$$\frac{0.024 \text{ mol}}{0.030 \text{ L}} = 0.80 \ M \text{ HCl}$$

Acids and Bases

Because pH is logarithmic, it takes 10x less base to move each step closer to 7. This creates a curve like this:

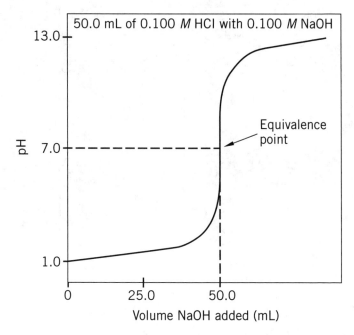

At first the pH changes slowly, but as the pH approaches 7, less NaOH is needed and the pH changes faster. After 7, the trend reverses itself.

For strong acid/strong base titration like this one, there is only water at equivalence, yielding a pH of 7.

$$H^+(aq) + OH^-(aq) \rightarrow H_2O(l)$$

Indicators 🛈

Indicators are used to signal the equivalence point of a titration by changing colors.

Indicators are weak acids that change colors around their pK_a value.

This occurs because the protonated and deprotonated forms of the indicator are different colors.

Indicators are often abbreviated using HIn. Phenolphthalein is an indicator which is colorless under a pH of 9, and pink above that.

Protonated		Deprotonated
HIn (aq)	⇌	In⁻(aq) + H⁺(aq)
Colorless		Pink

As weak acids, indicators usually start out mostly undissociated (so in their protonated form).

As OH⁻ is added to a titration, it reacts with HIn, creating more In⁻. When In⁻ > HIn, the color change is apparent.

When choosing an indicator, it's important to choose one that has a pK_a around the expected pH at equivalence.

Weak Acid/Strong Base Titrations !

The generic equation is HA(aq) + OH$^-$(aq) → A$^-$(aq) + H$_2$O(l).

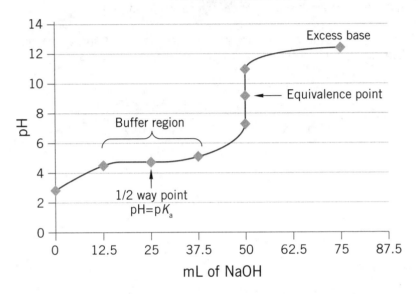

Important features:

- **Region 1:** Initial region. The steepness of the initial pH change is based on the K_a of the weak acid. Lower K_a = easier to change pH = bigger change.

- **Region 2:** Buffer region. As OH$^-$ is added, the amount of A$^-$ approaches the amount of HA, creating a buffer.

- **Region 3:** Equivalence region. No HA or OH$^-$ remain at equivalence, leaving only A$^-$. This reacts with water (A$^-$(aq) + H$_2$O(l) → HA(aq) + OH$^-$(aq)), creating a basic (pH > 7) equivalence point.

- **Region 4:** Excess base region. All weak acid has been deprotonated, leaving only excess OH$^-$ in solution.

At half-equivalence (which occurs halfway to the volume at equivalence), the pH of the solution = pK_a of the weak acid.

Weak Base/Strong Acid Titrations 🔔

The generic equation is B(aq) + H⁺(aq) → BH⁺(aq).

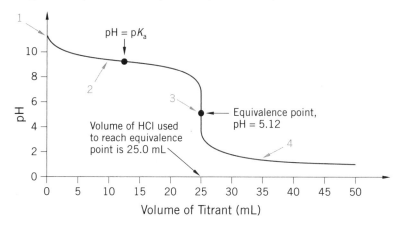

Important features on this diagram:

- **Region 1:** Initial region. The steepness of the initial pH change is based on the K_b of the weak acid. Lower K_b = easier to change pH = bigger change.

- **Region 2:** Buffer region. As H⁺ is added, the amount of BH⁺ approaches the amount of B, creating a buffer.

- **Region 3:** Equivalence region. No B or H⁺ remain at equivalence, leaving only BH⁺. This reacts with water (BH⁺(aq) + H₂O(l) → B(aq) + H₃O⁺(aq)), creating an acidic (pH < 7) equivalence point.

- **Region 4:** Excess acid region. All weak acid has been deprotonated, leaving only excess H⁺ in solution.

At half-equivalence, the pOH of the solution = pK_b of the weak acid.

Acids and Bases

Polyprotic Titrations 💀

In a polyprotic titration, the ratio between acids and bases is not 1:1. This must be accounted for in any calculations.

$$H_2CO_3(aq) + 2\ OH^-(aq) \rightleftharpoons CO_3^{2-}(aq) + 2\ H_2O(l)$$

Carbonic acid, H_2CO_3, is a diprotic acid. When 50.0 mL of H_2CO_3 of unknown concentration is titrated with some 0.50 M NaOH, the endpoint is reached after 45.50 mL of NaOH is added. What is the concentration of the carbonic acid?

$$(0.50\ M)(0.04550\ L) = 0.023\ \text{mol NaOH} \times \frac{1\ \text{mol}\ H_2CO_3}{2\ \text{mol NaOH}} =$$

$$\frac{0.012\ \text{mol}\ H_2CO_3}{0.050\ L} = 0.24\ M = [H_2CO_3]$$

You may see two regions of significant pH change in a polyprotic titration diagram. This is due to the multiple K_a values of a polyprotic acid.

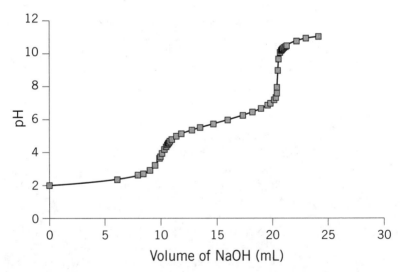

CHAPTER 8
Kinetics

It's now time to finally talk about *how quickly* reactions occur—we've spent the last several chapters discussing how to predict *if* reactions occur and to *what extent* they occur.

Factors Affecting Rate 🛈

In order for a reaction between two molecules to occur, they must collide with one another. The more collisions there are, the faster the reaction happens. This is the basis of what's known as **collision theory.** This theory states that when reactants collide with sufficient energy (known as the **activation energy, E_a**), a reaction occurs. The collisions that lead to a reaction are referred to as effective collisions.

Rate of Collisions

The rate of collisions will depend on two things: concentration and temperature.

1. As a function of concentration

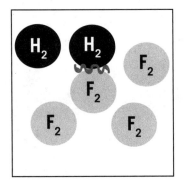 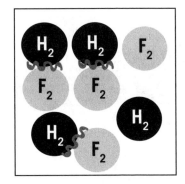

For gaseous or aqueous reactants, an increased concentration will increase the rate of reaction. This is because with more molecules moving around in a given volume, they are more likely to collide. For solid reactants, surface area plays a similar role to concentration.

2. As a function of temperature

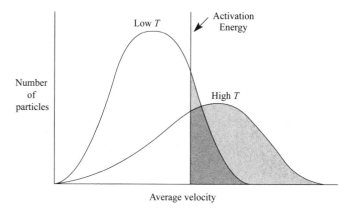

At a higher temperature, the reactant molecules are moving faster and will thus collide more often. In addition, a higher temperature means that a larger fraction molecules have a higher kinetic energy. Thus, at higher temperatures, a greater fraction of molecules will have sufficient collision energy to exceed the activation energy barrier.

Collision Orientation 🛑

In addition, reactions only occur if the reactants collide with the correct orientation. For example, in the reaction $2\ NO_2F \rightarrow 2\ NO_2 + F_2$, there are many possible collision orientations. Two of them are drawn on the following page.

Kinetics

Particulate Representations

[Diagram: FNO₂ + O₂NF → no reaction]

[Diagram: O₂NF + FNO₂ → NO₂ + F₂ + NO₂]

Activation Energy

As it turns out, the reaction coordinate diagrams we first discussed in the chapter on thermodynamics will be making a comeback here!

1. Endothermic vs. Exothermic Reaction Coordinates

    Given a reaction coordinate diagram, we can determine the E_a of the reaction by looking at the energy difference between the reactants and the highest energy point on the diagram:

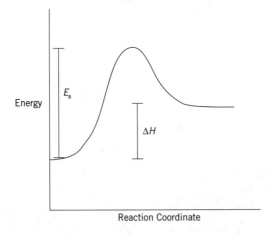

ENDOTHERMIC REACTION

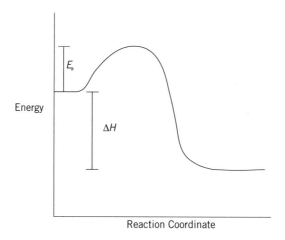

EXOTHERMIC REACTION

The applies for any reaction, whether it's exothermic or endothermic.

2. Forward vs. Reverse E_a

As we discussed in the chapter on equilibrium, reactions are reversible. The reverse reaction will have a different energy from the forward reaction, and it is the difference between the highest energy point on the diagram and the energy of the products:

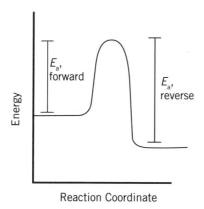

Rate of Disappearance/Appearance 🛑

The "rate of a reaction" is described in terms of the rate of appearance of a product or the rate of disappearance of a reactant. The rate of disappearance vs. appearance are related via coefficients.

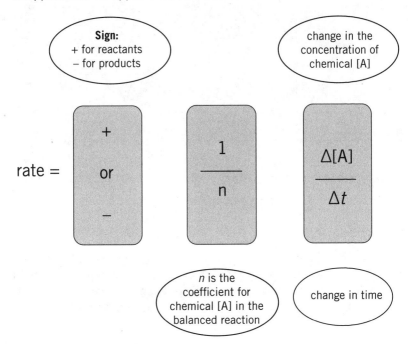

$$\text{rate} = \pm \, \frac{1}{n} \, \frac{\Delta[A]}{\Delta t}$$

- Sign: + for reactants, − for products
- n is the coefficient for chemical [A] in the balanced reaction
- change in the concentration of chemical [A]
- change in time

Here's an example of how we might measure this, using the reaction $2 H_2 + O_2 \rightarrow 2 H_2O$. Assume each sphere below = 1 mol of chemical and that the reaction occurs in a one liter flask.

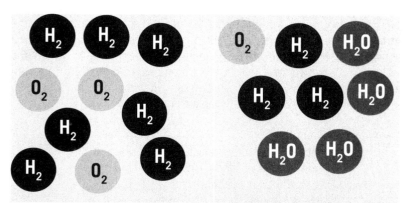

Initial system, at $t = 0$

System at $t = 1$ s

Four moles of H_2 have disappeared, so the change is $-\dfrac{1}{2}\dfrac{-4M}{s}$. We could also say that two moles of O_2 have disappeared: $-\dfrac{1}{1}\dfrac{-2M}{s}$. Or, four mols of H_2O have appeared: $+\dfrac{1}{2}\dfrac{4M}{s}$. These all simplify to the same reaction rate = $\dfrac{2M}{s}$.

Kinetics

Rate Laws 🛑

The **rate law** for a reaction describes the dependence of the initial rate of a reaction on the concentration of its reactants. It includes the **Arrhenius constant** (or **rate constant**), k, which takes into account the activation energy for the reaction and the temperature at which the reaction occurs. The rate law for a reaction cannot be determined from a balanced equation: it must be determined from experimental data, which is presented on the test in table form.

> **NOTE:** The rate constant depends on temperature and activation energy, but this will NOT be tested on the AP Chemistry Exam, so don't worry about it.

Reaction Orders

The rate law has the general form:

$$\text{rate} = k[A]^x[B]^y[C]^z$$

The greater the value of a reactant's exponent, the more a change in the concentration of that reactant will affect the rate of the reaction.

Zero/First/Second Order
- If the exponent is zero, doubling the concentration will have no effect on the rate. The rate is constant regardless of the concentration.
- If the exponent is one, doubling the concentration will double the rate.
- If the exponent is two, doubling the concentration will quadruple the rate.

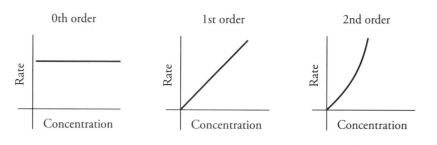

Calculations From Experimental Data

The easiest way to find the exponents in a rate law is to see what happens to the rate when the concentration of an individual reactant is doubled.

Expt.	[A] (M)	[B] (M)	Rate (M/s)
1	0.01	0.05	2.00×10^{-7}
2	0.01	0.10	4.00×10^{-7}
3	0.02	0.01	1.60×10^{-7}
4	0.04	0.01	6.40×10^{-7}

Comparing Runs 1 and 2 will allow us to find the order with respect to B:

Since the rate doubles when [B] is doubled, the reaction is first-order in B.

Comparing Runs 3 and 4 allow us to find the with respect to A:

the rate quadruples [A] is doubled, the ion is second-order

rate = $k[A]^2[B]^1$

Orders for Individual reactants vs. Overall Reaction Order

A reaction with the rate law rate = $k[A]^2[B]$ is said to be second-order with respect to A and first-order with respect to B. The overall order of the reaction is given by adding up the orders of the individual reactants, so in this case the reaction is third-order overall.

Rate Constant

Once the rate law has been determined, the value of the rate constant, k, can be calculated using any of the lines of data in the table. Let's use experiment 1.

$$k = \frac{\text{rate}}{[A]^2[B]} = \frac{2 \times 10^{-7} \text{ M/s}}{(0.01 \text{ M})^2 (0.05 \text{ M})} = 0.04 \left(\frac{\text{M/s}}{\text{M}^3}\right) = 0.04 \frac{1}{\text{M}^2 \text{s}}$$

The units of k depend on the overall order of the reaction and can be figured out with dimensional analysis.

Integrated Rate Laws

It's also useful to have rate laws that relate the rate constant k to the way that concentrations change over time. These graphs can also be used to determine the value of the rate constant. To do this, we make three graphs to see which gives us the straightest line: [A] vs. time, ln[A] vs. time, and 1/[A] vs. time.

- Zero order: [A] vs. time

    The rate of a zero-order reaction does not depend on the concentration of reactants at all, so the graph of the change in concentration of a reactant of a zero-order reaction versus time will be a straight line with slope equal to $-k$.

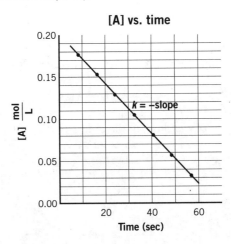

- First order: ln[A] vs. time

    The rate law for a first-order reaction uses natural logarithms. Therefore, plotting ln[A] vs time will create a linear graph with slope $-k$ and y-intercept $\ln[A]_0$.

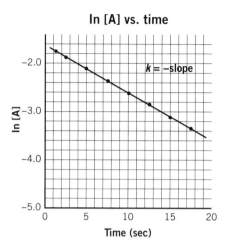

Using slope-intercept form, we can interpret this graph to come up with a useful equation:

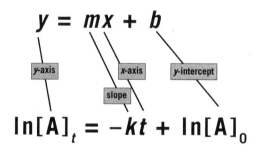

$$\ln[A]_t = -kt + \ln[A]_0$$

$[A]_t$ = concentration of reactant A at time t
$[A]_0$ = initial concentration of reactant A
k = rate constant
t = time elapsed

Kinetics

- Second order: 1/[A] vs. time

    The rate law for a second-order reaction uses the inverses of the concentrations.

    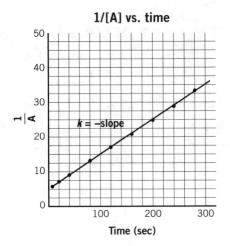

    $$y = mx + b$$

    y-axis: $\dfrac{1}{[A]_t}$

    slope: k

    x-axis: t

    y-intercept: $\dfrac{1}{[A]_0}$

    $$\frac{1}{[A]_t} = kt + \frac{1}{[A]_0}$$

First order = Constant Half-Life

The half-life of a reactant in a chemical reaction is the time it takes for half of the substance to react. Most half-life problems can be solved by using a simple chart. Don't go half way on memorizing this half-life chart (sorry, couldn't resist).

Time	Sample	Time (half-life = 3 y)	Sample
0	100%	0	120 g
1 half-life	50%	3 years	60 g
2 half-lives	25%	6 years	30 g
3 half-lives	12.5%	9 years	15 g

For a first-order reactant, half-life remains constant. This can be demonstrated graphically as shown below:

$$\text{half} - \text{life} = \frac{\ln 2}{k} = \frac{0.693}{k}$$

Decay of a Radioactive Substance

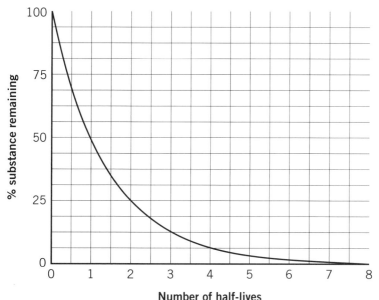

Kinetics

Note that for zero-order or second-order reactants, half-life is NOT constant. You cannot use the half-life equation for anything other than a first-order reactant.

One application of half-life is to examine the rate of decay of a radioactive substance. A radioactive substance is one that will slowly decay into a more stable form as time goes on. (In chemistry, unlike in one's personal life, slow decay leads to stability. Hey, I don't make the rules.)

Experimental Design

Beer's Law (You Don't Have To Be 21 Though)

Now let's get this party started by talking about Beer's Law. While it has nothing to do with beer, it is about how substances are absorbed. To measure the concentration of a solution over time, a device called a spectrophotometer can be used in some situations. A **spectrophotometer** measures the amount of light at a given wavelength that is absorbed by a solution. If a solution changes color as the reaction progresses, the amount of light that is absorbed will change.

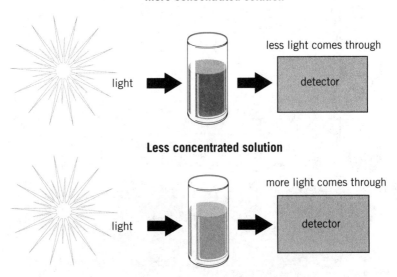

Beer's Law

$$A = abc$$

A = absorbance
a = molar absorptivity, a constant that depends on the solution
b = path length, the distance the light is travelling through the solution
c = concentration of the solution

Since molar absorptivity and path length are constants when using a spectrophotometer, Beer's Law is often interpreted as a direct relationship between absorbance and the concentration of the solution.

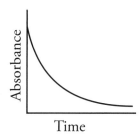

Since absorbance and concentration are directly proportional, a plot of ln[absorbance] vs. time should produce a straight line with slope $-k$ if the reaction is first-order.

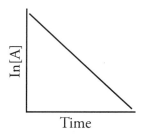

Reaction Mechanisms

Many chemical reactions are not one-step processes. Rather, the balanced equation is a sum of series of individual steps, called **elementary steps.**

Example mechanism

$$2\ NO_2Cl \rightarrow 2\ NO_2 + Cl_2$$

Step 1. $NO_2Cl \rightarrow NO_2 + Cl$ (slow)   ← this elementary step has only one reactant; it's unimolecular

Step 2. $NO_2Cl + Cl \rightarrow NO_2 + Cl_2$ (fast)   ← this elementary step has two reactants; it's bimolecular

---

In this reaction, Cl appears in the elementary steps, but not in the overall balanced reaction. Cl is a reaction **intermediate:** it is produced but also fully consumed over the course of the reaction. Intermediates will always cancel out when adding up the various elementary steps in a reaction.

Coefficients = Order for Elementary Steps

The rate law for an elementary step can be determined by taking the concentration of the reactants in that step and raising them to the power of any coefficient attached to that reactant.

Step 1. $NO_2Cl \rightarrow NO_2 + Cl$ (slow)   rate = $k[NO_2Cl]$

Step 2. $NO_2Cl + Cl \rightarrow NO_2 + Cl_2$ (fast)   rate = $k[NO_2Cl][Cl]$

Multi-Step Reaction Coordinates

We can think of each elementary step as its own mini-reaction, with a reaction energy diagram to match which shows the energy level of the products and reactants, as well as the required activation energy for a reaction to occur.

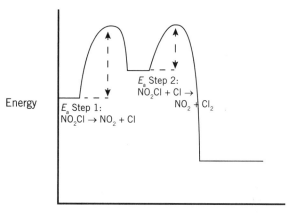

We can see that the first step is endothermic and the second step is exothermic, as well as that the overall reaction is exothermic.

Determining Rate Laws

We can use reaction mechanisms to determine rate laws.

The rate law for the entire reaction is equal to that of the slowest elementary step (also called the rate-determining step), which is Step 1 in the mechanism we've been looking at.

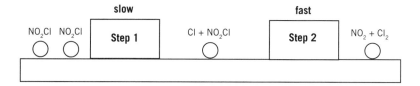

Like a factory conveyor belt, products cannot be produced any faster than the slowest step.

The slow step is not always the first one, though. Let's take a look at a different mechanism:

Rate law for a reaction with a slow step preceded by a fast equilibrium step

Step 1. $2A \rightleftharpoons X$ Rate = $k[A]^2$ (fast)

Step 2. $X + B \rightarrow Y$ Rate = $k[X][B]$ (slow)

Step 3. $Y + B \rightarrow D$ Rate = $k[Y][B]$ (fast)

The rate law would be expected to be rate = $k[X][B]$ from this mechanism. However, step 2 has an intermediate (X) present in it. Even though intermediates can appear in rate laws, if they can be replaced with a reactant via an earlier step, we should try and do so. Looking at step 1, we also see that [X] is equivalent to $[A]^2$ since the two sides are in equilibrium.

Catalysts and Intermediates

Intermediates are the chemicals in a mechanism which are produced in one step and used in another. They don't appear in the balanced equation. Neither do catalysts; these chemicals are used up in one step but then regenerated in a later step.

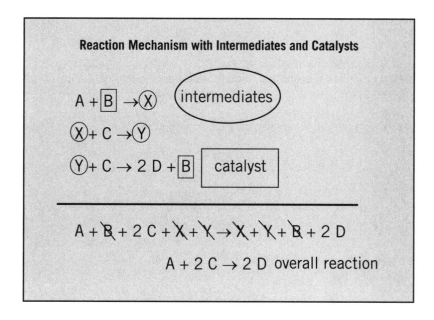

Catalysts increase the rate of reaction by providing an alternative reaction mechanism that has a lower activation energy.

Types of Catalysis: Acid-base and Surface 💀

You may be required to state the mechanisms by which catalysts work: two common ones are acid-base catalysis and surface catalysis.

Types of Catalysis	
Acid-base Catalysis	Surface Catalysis
An acid may react with one of the reactants to make it more reactive. $$CH_3OH + CH_3COOH \xrightarrow{H^+} CH_3OOCH_3 + H_2O$$	The metal surface may react with one of the reactants, making it more reactive. $$N_2 + 3H_2 \xrightarrow{Fe} 2NH_3$$

Kinetics

Catalysts

A catalyst lowers the activation energy for a reaction, and this can be seen on a reaction coordinate diagram when the energy diagrams for the catalyzed and uncatalyzed reactions are compared.

Here's an example of catalysts on reaction coordinates:

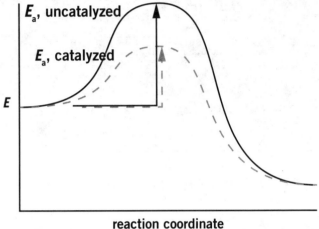

Remember that catalysts may make an appearance in rate laws, if they are involved in the rate determining step.

One important type of catalyst that you should know is a type that is present in biological systems. These biological catalysts are called **enzymes** and they often work by bringing together reactants in the correct orientation or providing some alternative pathway for the reaction which requires less activation energy.

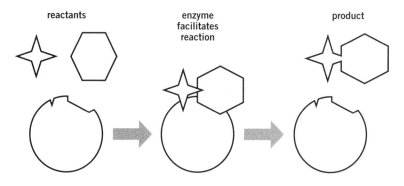

Turn the page for a handy Periodic Table and a tally of AP Chemistry Equations and Constants. You're welcome!

PERIODIC TABLE OF THE ELEMENTS

DO NOT DETACH FROM BOOK.

1 H 1.008																	2 He 4.00
3 Li 6.94	4 Be 9.01											5 B 10.81	6 C 12.01	7 N 14.01	8 O 16.00	9 F 19.00	10 Ne 20.18
11 Na 22.99	12 Mg 24.30											13 Al 26.98	14 Si 28.09	15 P 30.97	16 S 32.06	17 Cl 35.45	18 Ar 39.95
19 K 39.10	20 Ca 40.08	21 Sc 44.96	22 Ti 47.90	23 V 50.94	24 Cr 52.00	25 Mn 54.94	26 Fe 55.85	27 Co 58.93	28 Ni 58.69	29 Cu 63.55	30 Zn 65.39	31 Ga 69.72	32 Ge 72.59	33 As 74.92	34 Se 78.96	35 Br 79.90	36 Kr 83.80
37 Rb 85.47	38 Sr 87.62	39 Y 88.91	40 Zr 91.22	41 Nb 92.91	42 Mo 95.94	43 Tc (98)	44 Ru 101.1	45 Rh 102.91	46 Pd 106.42	47 Ag 107.87	48 Cd 112.41	49 In 114.82	50 Sn 118.71	51 Sb 121.75	52 Te 127.60	53 I 126.91	54 Xe 131.29
55 Cs 132.91	56 Ba 137.33	57 *La 138.91	72 Hf 178.49	73 Ta 180.95	74 W 183.85	75 Re 186.21	76 Os 190.2	77 Ir 192.2	78 Pt 195.08	79 Au 196.97	80 Hg 200.59	81 Tl 204.38	82 Pb 207.2	83 Bi 208.98	84 Po (209)	85 At (210)	86 Rn (222)
87 Fr (223)	88 Ra 226.02	89 †Ac 227.03	104 Rf (261)	105 Db (262)	106 Sg (266)	107 Bh (264)	108 Hs (277)	109 Mt (268)	110 Ds (271)	111 Rg (272)							

*Lanthanide Series

58 Ce 140.12	59 Pr 140.91	60 Nd 144.24	61 Pm (145)	62 Sm 150.4	63 Eu 151.97	64 Gd 157.25	65 Tb 158.93	66 Dy 162.50	67 Ho 164.93	68 Er 167.26	69 Tm 168.93	70 Yb 173.04	71 Lu 174.97

†Actinide Series

90 Th 232.04	91 Pa 231.04	92 U 238.03	93 Np (237)	94 Pu (244)	95 Am (243)	96 Cm (247)	97 Bk (247)	98 Cf (251)	99 Es (252)	100 Fm (257)	101 Md (258)	102 No (259)	103 Lr (262)

ADVANCED PLACEMENT CHEMISTRY EQUATIONS AND CONSTANTS

Throughout the test the following symbols have the definitions specified unless otherwise noted.

L, mL	= liter(s), milliliter(s)	mm Hg	= millimeters of mercury
g	= gram(s)	J, kJ	= joule(s), kilojoule(s)
nm	= nanometer(s)	V	= volt(s)
atm	= atmosphere(s)	mol	= mole(s)

ATOMIC STRUCTURE

$E = h\nu$
$c = \lambda\nu$

E = energy
ν = frequency
λ = wavelength

Planck's constant, $h = 6.626 \times 10^{-34}$ J s
Speed of light, $c = 2.998 \times 10^8$ m s^{-1}
Avogadro's number = 6.022×10^{23} mol^{-1}
Electron charge, $e = -1.602 \times 10^{-19}$ coulomb

EQUILIBRIUM

$K_c = \dfrac{[C]^c[D]^d}{[A]^a[B]^b}$, where $aA + bB \rightleftarrows cC + dD$

$K_p = \dfrac{(P_C)^c(P_D)^d}{(P_A)^a(P_B)^b}$

$K_a = \dfrac{[H^+][A^-]}{[HA]}$

$K_b = \dfrac{[OH^-][HB^+]}{[B]}$

$K_w = [H^+][OH^-] = 1.0 \times 10^{-14}$ at 25°C
$ = K_a \times K_b$

pH = $-\log[H^+]$, pOH = $-\log[OH^-]$

$14 = $ pH + pOH

pH = $pK_a + \log\dfrac{[A^-]}{[HA]}$

$pK_a = -\log K_a$, $pK_b = -\log K_b$

Equilibrium Constants
K_c (molar concentrations)
K_p (gas pressures)
K_a (weak acid)
K_b (weak base)
K_w (water)

KINETICS

$\ln[A]_t - \ln[A]_0 = -kt$

$\dfrac{1}{[A]_t} - \dfrac{1}{[A]_0} = kt$

$t_{1/2} = \dfrac{0.693}{k}$

k = rate constant
t = time
$t_{1/2}$ = half-life

NOTES

NOTES